T0135523

Bibliografische Information der Deutschen Nationalbibliothek

Die Deutsche Nationalbibliothek verzeichnet diese Publikation in der
Deutschen Nationalbibliografie; detaillierte bibliografische Daten sind
im Internet über http://dnb.d-nb.de abrufbar.

ISBN 978-3-8325-3850-7

Logos Verlag Berlin GmbH
Comeniushof, Gubener Str. 47,
10243 Berlin
Tel.: +49 (0)30 42 85 10 90
Fax: +49 (0)30 42 85 10 92
INTERNET: http://www.logos-verlag.de

Data Compression and Compressed Sensing in Imaging Mass Spectrometry and Sporadic Communication

von Andreas Bartels

Dissertation

zur Erlangung des Grades eines Doktors der Naturwissenschaften
— Dr. rer. nat. —

Vorgelegt im Fachbereich 3 (Mathematik & Informatik)
der Universität Bremen
im August 2014

Datum des Promotionskolloquiums:
25. September 2014

Gutachter:
Dr. Dennis Trede (Universität Bremen)
Prof. Dr. Dirk A. Lorenz (Technische Universität Braunschweig)

Fassung:
Überarbeitete Fassung im Umfang der Anmerkungen der Gutachten.

Data Compression and Compressed Sensing

in Imaging Mass Spectrometry and Sporadic Communication

—————

Andreas Bartels

Abstract

This thesis contributes to the fields of data compression and compressed sensing (CS) and their application to imaging mass spectrometry (IMS) and sporadic communication. Since the pioneering work by Candès, Romberg, Tao and Donoho between 2004 and 2006 it is clear that it is possible to compress data during the acquisition process instead of first measuring large data and compressing it afterwards. CS is mainly built on the knowledge that most data is compressible or sparse, meaning that most of its content is redundant and not worth being measured.

The first part gives an introduction to the basic notation and mathematical concepts that are used. Mathematical compression techniques and a detailed overview on CS principles are presented as well.

The second and the third part build the main parts of the thesis. They deal with IMS data and possible compression methods. Numerical results on a rat brain dataset and rat kidney dataset complete both parts.

In the second part, a new and more sensitive peak picking method is introduced. It considers the spatial information of each m/z-image in the data by using a new measure which evaluates the respective structuredness. This enables automatic subdivision in structured and unstructured images.

In the third part, a first CS model for IMS is introduced. It combines peak-picking of the spectra and denoising of the m/z-image which are both typical post-processing steps. A robustness result for the reconstruction of compressed measured data is presented. It generalizes known reconstruction guarantees for the ℓ_1- and TV-cases by taking into account both sparsity aspects in the hyperspectral data. The compressed sensing model and the robustness result build the main results of the thesis.

Finally, in the fourth part, the topic of CS based multi-user detection in sporadic communication is discussed. For reconstruction of the transmitted signals, the elastic-net functional is considered for which a new parameter choice rule similar to the L-curve method is presented. It is applicable in both an online (i.e. in real-time) and offline (i.e. calculated in advance) scenario in the considered application.

Zusammenfassung

Diese Arbeit ist ein Beitrag in den Bereichen der Kompressionverfahren und des Compressed Sensing (CS) mit den Anwendungsschwerpunkten der bildgebenden Massenspektrometrie (MS) und der sporadischen Kommunikation. Seit den Pionierarbeiten von Candès, Romberg, Tao und Donoho ist bekannt, dass es möglich ist, Daten bereits während ihrer Erhebung und nicht erst nach ihrer vollständigen Messung zu komprimieren. CS basiert in erster Linie auf der Annahme, dass die zugrundeliegenden Daten komprimierbar oder sparse, d.h. redundant sind.

Im ersten Abschnitt wird eine Einführung in die Notation und die verwendeten Grundlagen für die nachfolgenden Kapitel gegeben. Darüber hinaus werden neben bekannten Kompressionstechniken auch Grundideen und wesentlichen Resultate des CS vorgestellt.

Im zweiten Abschnitt wird eine neue und sensiblere Peak-Picking-Methode für die bildgebende Massenspektrometrie vorgestellt. Anders als bekannte in spektraler Dimension orientierte Verfahren wird dabei die räumliche Perspektive im Bezug auf jedes einzelne m/z-Bild eingenommen. Ein neu definiertes Maß bewertet für jedes dieser Bilder den Grad der Strukturiertheit, was eine automatische Unterteilung in strukturierte und unstrukturierte Bilder ermöglicht.

Im dritten Abschnitt wird ein erstes CS-Modell für die Anwendung in der bildgebenden MS vorgestellt. Dieses kombiniert die sonst üblichen Nachbearbeitungsprozeduren des Peak-Picking auf den Spektren und des Entrauschens der einzelnen Bilder. Es wird nachgewiesen das die Rekonstruktion der gemessenen Daten durch das vorgeschlagene Minimierungsproblem unter gewissen Annahmen an die zugrundeliegenden Operatoren robust gegenüber Rauschen ist. Dieses Resultat vereint bereits bekannte ähnliche Resultate für die ℓ_1- and TV-Fälle und erweitert diese auf hyperspektrale Daten. Das CS-Modell und das Robustheitsresultat bilden die Hauptresultate dieser Arbeit.

Im vierten und letzten Abschnitt wird ein bekanntes Modell für die CS-basierte Mehrbenutzer-Erkennung in der sporadischen Kommunikation dargestellt. Für die Rekonstruktion übertragener Signale wird das Elastic-Net-Funktional verwendet, wofür eine neue Parameterwahlstrategie basierend auf der bekannten L-Kurve vorgestellt wird.

Acknowledgments

I thank Dr. Dennis Trede and Prof. Dr. Peter Maaß for their persistent supervision, support and trust throughout my studies. They not only gave me the chance to work in a very interesting field of applied mathematics, but also to be part in different intriguing projects.

I am indebted to Patrick Dülk, Jan Hendrik Kobarg, Thomas Page, Dr. Theodore Alexandrov and Henning Schepker for advice and fruitful exchange of ideas. Moreover, I thank Patrick Dülk, Dr. Jan Hendrik Kobarg, Andrew Palmer, David Rea and Dr. Carsten Bockelmann for careful reading of the manuscript and for giving helpful hints and suggestions. I want to thank all my colleagues at "Zentrum für Technomathematik" for the good time and the pleasant working atmosphere.

I acknowledge the financial support from the BMBF project HYPER-MATH "Hyperspectral Imaging: Mathematische Methoden für Innovationen in Medizin und Industrie" under the grant 05M13LBA.

Finally I would like to thank my girlfriend Henrike, my family, but especially my mum for everything she has done in her life to make this possible.

Andreas Bartels
Zentrum für Technomathematik
Universität Bremen

ix

x

Contents

Contents

1 | Introduction

This thesis is devoted to two different but parallel fields in both the pure and applied mathematics, namely *data compression* and *compressed sensing*. Several techniques are presented with the main focus on the application to imaging mass spectrometry. In addition, also a compressed sensing based approach is treated for multi-user detection in sporadic communication. The contributions of this thesis are described in detail at the end of the introduction.

1.1. The big data problem

In 2010, the news magazine *The Economist* published a special report on managing information entitled "Data, data everywhere" [185]. As the name already indicates, this special issue deals with a rising problem of the huge amount of data that is collected and the difficulties to store, interpret, or in more general, *deal* with them. Interviewed for this report, the *The Economists'* data editor Kenneth Neil Cukier mentions that "information has gone from scarce to superabundant" and therefore "brings huge new benefits – but also big headaches". Among all the different reasons for the data explosion mentioned, it is probably the enhancement of technology which has lead to this situation, well summarized in the observation: "As the capabilities of digital devices soar and prices plummet, sensors and gadgets are digitizing lots of information that was previously unavailable." [185]. With the decreasing prices for mobile- or smart phones, for example, the number of their users increases alongside an according rate of information exchanged.

There are several associated problems with *big data*. First, depending on the acquisition method used, the time for collecting the data can be long, as for example in magnetic resonance imaging (MRI) [132, 133]. In addition, an interpretation of big data might be difficult since an automatic extraction of its main features is usually missing. Finally, the data to store might be inadequately large compared to its content, which makes a subsequent *compression*, i.e. a reasonable reduction of the data size, necessary. The latter is a still common step in modern acquisition systems such as in photography. There, a photo is taken by a certain number of image sensors which is usually called the number of pixels of the camera. After that, the image is compressed with only little visible degradation, probably in the most common file format, the *Joint Photography Experts Group* standard (JPEG) [156]. This simply allows many more images to be saved in the camera memory than one could without compression. This example clearly demonstrates one big data problem: On the one hand one has the technique to measure lot of data, but on the other hand one does not have that much memory to save it all. The next section shortly reviews some of the common compression techniques.

1.2. Data compression

As already motivated in the previous section, compression techniques are indispensable in the modern time in which huge amounts of data are collected. One generally differentiates between *lossless* and *lossy compression* techniques. Lossless compression means that it is possible to reconstruct the data to its original state after compression. In contrast, lossy compression deliberately avoids a perfect recoverability, resulting in a higher compression rate. This, however, does not necessarily mean that the loss of detail is actually visible with the human eye, as the following example will show. Amongst many other data, photographs have been found to be extremely compressible. The above mentioned JPEG image format[1], for example, uses the discrete cosine transform [5] to achieve compression rates of about 10:1 [156]. This rate could be improved via the JPEG 2000 format [184] which applies the discrete wavelet transformation [65, 131, 134]. Other prominent examples for almost daily data

---

[1]The JPEG compression scheme is principally able to do either lossy or lossless compression, but is mostly used as in the first mentioned case [164].

formats are the lossy audio compression scheme MPEG Audio Layer III (MP3) or the lossy video compression scheme MPEG from the *Moving Picture Experts Group* with all its derivates incurred in the past [164]. What all approaches have in common is the fact that they take advantage of the compressibility of the data, meaning that it suffices to take only very few cosine or wavelet basis functions for an adequate representation of the data.

All compression strategies have in common that they need the full data to work with, even though most of it is redundant. Therefore, it is natural to ask if the measuring or sensing as well as the compression step could be combined, which would be also useful for acquisition systems where sensors are expensive, limited or slow [50]. Fortunately, and this is one of the main concepts within this thesis, the question can be positively answered via a topic called *compressed sensing*.

1.3. What compressed sensing is about

Since the pioneering work of Candès, Romberg, Tao and Donoho in 2006 [46, 47, 49, 72], the field of *compressed sensing, compressive sensing* or *compressive sampling* (CS) arises to be more and more attractive to study, as this states a non-intuitive counterpart of the known Nyquist-Shannon sampling theorem [152, 175]. Instead of measuring at least twice the maximum frequency of the signal for a proper reconstruction [195], it astonishingly turns out that only few nonadaptive randomly sampled information are sufficient for exact or near-optimal signal recovery under certain conditions. This groundbreaking insight has rapidly inspired a tremendous effort in studying the theoretical but also the practical areas in this field.

CS is built on two key principles. The first is that many signals have the property to be *sparse* or *compressible* in a proper basis Ψ, which means that most parts of them are negligible. The second is that the sensing waveforms given via a sensing matrix Φ have a dense representation in Ψ, called low *(mutual) coherence*. This can also be interpreted in the way that the smallest angle between the rows $\bar{\varphi}_i$ of Φ and the columns of ψ_j of Ψ is large, which implies that the differentiation between taken measurements and the sensed signal in the basis Ψ is still feasible. Given a signal x that is known to be sparse with respect to Ψ, it is natural

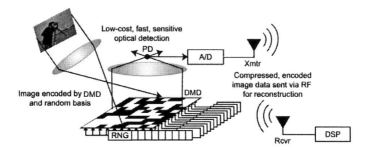

Figure 1.1.: Single-pixel camera, taken from [197], © 2006, IEEE.

to ask how the sensing matrix Φ should be constructed to reach a low coherence. Surprisingly, constructing sensing matrices randomly leads to low coherence for a wide range of basis matrices with high probability. Randomness plays a key role in CS and with that concept established, the question arises whether this is realizable in practice. In fact, acquisition devices usually follow a deterministic routine such as in a camera where each pixel of a given scene is measured step by step or simultaneously, but not at random. With the foundations laid by the theoretical success of CS, work was and is done to transfer theory to practical pseudo random measurement systems.

One prominent example is given by the single-pixel camera [197] which implements the random measurements via a Digital Micromirror Device (DMD), see Figure 1.1. The DMD is, in simple terms, nothing else than a matrix comprising of tiny mirrors which can be closed or open. The state of each can be set at random. However, instead of taking a full screen measurement by acquiring each pixel of the scenery separately, on aims in taking only few samples, i.e. less than the numbers of image pixels. Baraniuk, Duarte *et al.* [79, 183, 197] have shown in examples that it is possible to reconstruct an image well from up to 10% random measurements, acquired via the single-pixel camera. To illustrate the ideas of CS, especially the sparsity and randomness aspects, it will now follows a nice *combinatorial group test* example that has been published by Bryan and Leise in an introductory article on CS [37].

Suppose it is given the task to find out the one coin among given seven gold coins $c_1, \ldots, c_7$ that has a different mass and therefore constitutes a forgery. Moreover, one is limited to use an accurate electronic scale at most three times. In the case the mass of a true coin is known, one could first weigh the coins c_1, c_3, c_5, c_7 together. For the second weighing the coins c_2, c_3, c_6, c_7 and for the last c_4, c_5, c_6, c_7 would be put on the scale. This taken selection can also be expressed via the following 0-1 matrix Φ whose k-th row corresponds to the k-th weighing:

$$\Phi = \begin{pmatrix} 1 & 0 & 1 & 0 & 1 & 0 & 1 \\ 0 & 1 & 1 & 0 & 0 & 1 & 1 \\ 0 & 0 & 0 & 1 & 1 & 1 & 1 \end{pmatrix}. \tag{1.1}$$

It is observable that the k-th column corresponds to the k-th integer in binary if the k-th row corresponds to 2^{k-1}. Also note that the sensing matrix in (1.1) is therefore structured what will later on be contrasted to the random matrices. In addition, it is clear that the identification of the forgery is unique by these three weighings. If there is no wrong coin within the seven one would all three times weigh four times the mass of a single coin. This clearly changes if one coin is wrong. As an example, assume that one detects a difference in the 3rd measurement, then, coin c_4 is the counterfeit. Otherwise, say, if the first and the third measurements are different, one deduces that the coin c_5 is wrong. However, in a general setting of n coins it could follow the same technique and do around $m = \log_2(n)$ measurements to detect the forgery.

Now consider the case of $n = 100$ coins including three counterfeits. More formally, let x_i of $x \in \mathbb{R}^n$ denote the deviation of the i-th coin from its nominal mass. Exemplarily say that only the coins 14, 40 and 73 are counterfeits with deviations $x_{14} = 0.32$, $x_{40} = -0.35$ and $x_{73} = 0.25$, i.e. $x_i = 0$ for all the other ones, see Figure 1.2(a). The question is how to detect those three bad coins by only a very small number $m \ll n$ of weighings? One could probably think of the same approach as above with a $m \times n$ 0-1 matrix. But unfortunately there is more than one counterfeit and the binary encoding strategy might fail.

What is now coming into play is the counterintuitive part of CS; one takes each of the m measurements at *random*, i.e. each entry is independently set as either zero or one with equal probability. In this example, the aim is to detect the three forgeries from $m = 20$ measurements of the

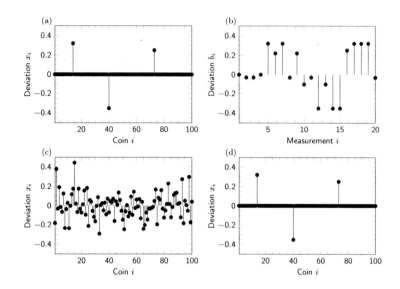

Figure 1.2.: Visualization of the coin example and used regularization approaches. (a) presents the sparseness of the deviation vector x as only three coins are forgeries (x_{14}, x_{40} and x_{73}), (b) shows the compressed sensing measurements b. (c) and (d) present the reconstruction results based on ℓ_2- and ℓ_1-regularization, respectively.

$n = 100$ given coins. More formally, let $x \in \mathbb{R}^{100}$ be the deviation vector and $\bar{\varphi}_k \in \{0, 1\}^{100}$, $k = 1, \ldots, 20$, the k-th measurement vector given as the k-th row of $\Phi = (\Phi_{k,i}) \in \{0, 1\}^{20 \times 100}$. Then the deviation of the k-th measured subset $b_k \in \mathbb{R}$, visualized in Figure 1.2(b), can be written as

$$b_k = \langle \bar{\varphi}_k, x \rangle = \bar{\varphi}_k^T x = \sum_{i=1}^{100} \Phi_{k,i} x_i.$$

In matrix form with $b = (b_1, \ldots, b_{20})^T$ and $\Phi = (\bar{\varphi}_1, \ldots, \bar{\varphi}_{20})^T$ this becomes

$$\Phi x = b. \tag{1.2}$$

Put now the focus on the last equation (1.2) and remember that the goal is still to find the counterfeits, i.e. those coins with deviations in terms of the mass within the given 100 ones. That means that x is unknown and it is only given the measurement system Φ as well as the resulting subset deviations in b. As there are 20 taken measurements but 100 unknowns, this system is highly underdetermined, resulting in infinitely many possible solutions.

An often used technique is to take that solution x° to $\Phi x = b$ which has minimum Euclidean ℓ_2-norm $\|x\|_2 = (\sum_i x_i^2)^{1/2}$, i.e.

$$\operatorname*{argmin}_x \|x\|_2 \quad \text{subject to} \quad \Phi x = b. \tag{1.3}$$

This optimization problem can be solved by, for example, Lagrange multipliers. For this one usually recasts problem (1.3) in the following equivalent formulation

$$\operatorname*{argmin}_x \|x\|_2^2 \quad \text{subject to} \quad \Phi x = b. \tag{1.4}$$

The Lagrange function then reads $L(x, \tau) = \|x\|_2^2 + \tau^T(\Phi x - b)$ with Lagrange multiplier τ. The optimality conditions with respect to both unknowns x and τ then directly lead to $2x^\circ + \Phi^T \tau^\circ = 0$ as well as $\Phi x^\circ = b$, and therefore the explicit minimum norm solution $x^\circ = \Phi^T (\Phi \Phi^T)^{-1} b$. Unfortunately, and this will be elucidated in more detail in Chapter 3, it turns out that in many cases the ℓ_2-norm is not an appropriate choice as this tends to solutions which are almost never sparse. As shown in Figure 1.2(c), the ℓ_2-norm solution x° to the coin problem is indeed non-sparse and far away from the truth. Nevertheless, it needs to be mentioned that this x° satisfies (1.2) exactly.

A common step is now to impose more appropriate *a priori* information on the unknown x. To exploit the a priori known sparsity of the deviation vector x, one could think of a norm that simply counts the number of non-zero entries and which is denoted by $\| \cdot \|_0$. For a vector $x \in \mathbb{R}^n$ it formally reads

$$\|x\|_0 := |\{i \mid x_i \neq 0\}|.$$

This expression is often named ℓ_0-"norm" even though it is in fact not a norm since the absolute homogeneity is not fulfilled. This norm tends to be a right choice, but regrettable leads to an NP-hard problem [148].

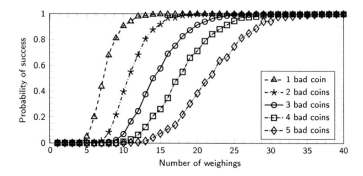

Figure 1.3.: Success rates for the coin example and ℓ_1-regularization. 1000 repetitions were performed with $n = 100$ coins and changing number of measurements, $m \in \{1, \ldots, 40\}$, and faulty coins, $s \in \{1, \ldots, 5\}$.

Fortunately, it turns out that the ℓ_1-norm $\|x\|_1 = \sum_i |x_i|$ is a good choice for remaining both computational tractability but also sparseness of the desired solution:

$$\operatorname*{argmin}_x \|x\|_1 \quad \text{subject to} \quad \Phi x = b. \tag{1.5}$$

If one incorporates possible noise w on the measurements, i.e. $\Phi x + w = b$, $\|w\|_2 \leqslant \varepsilon$ where $\varepsilon > 0$, then (1.5) becomes

$$\operatorname*{argmin}_x \|x\|_1 \quad \text{subject to} \quad \|\Phi x - b\|_2 \leqslant \varepsilon. \tag{1.6}$$

Especially with the beginning and success of CS, ℓ_1-minimization problems or *basis pursuit* [56] as in (1.5) or *quadratically constrained basis pursuit* as in (1.6) gained much attention [22, 24, 27–29, 40, 59, 93, 149]. One possible way to solve problem (1.6) is to rewrite it to the following Lagrangian or Tikhonov type functional [66],

$$\operatorname*{argmin}_x \|x\|_1 + \frac{\lambda}{2} \|\Phi x - b\|_2^2. \tag{1.7}$$

Note that both problems (1.6) and (1.7) are equivalent for some proper choice of λ and ε, respectively [93, Proposition 3.2]. In a similar way as for the ℓ_2-problem the optimality condition for a potential minimizer can be formulated. This leads to the iterative soft-thresholding algorithm

$$x_{k+1} = \mathbb{S}_\gamma(x_k - \gamma\lambda\Phi^T(\Phi x_k - b)) \qquad (1.8)$$

for finding a solution x to (1.5). The notation $\mathbb{S}_\gamma(\cdot)$ describes the soft-thresholding or shrinkage operator with threshold γ (see also Section 2) which is in component notation given as

$$[\mathbb{S}_\gamma(x)]_i = \max\{0, |x_i| - \gamma\}\operatorname{sign}(x_i). \qquad (1.9)$$

It was shown in [35] that the iteration (1.8) converges linearly to the minimizer of (1.7). Tremendous efforts have been made to find fast and accurate algorithms to solve this or similar kinds of problems. To mention only a few, there are NESTA [24], FISTA [22], GPSR [90], SL0 [143] or even whole program packages such as ℓ_1-MAGIC [44].

In any case, calculating a minimizer via (1.8) in fact leads to the *exact*[2] sparse solution as shown in Figure 1.2(d). But were the 20 randomly chosen measurements given in Φ just luckily chosen? To verify this, it is now carried out the experiment with three bad coins 1000 times and different number of weighings; see red full circled line in Figure 1.3. In total one obtains 978 exact reconstructions. If the number of measurements is slightly increased to $m = 25$, one always gets exact solutions in the experiment. Conversely, if the number of weighings is decreased to $m = 5$, only 289 successes are recognizable. In addition, corresponding curves for 1, 2, 4 and 5 counterfeit coins are also shown in Figure 1.3. As one can see, the less sparse the deviation vector x is, the more measurements are usually needed for perfect reconstruction.

This example, especially the presentation of the interaction of the number of taken measurements m and the sparsity level s in Figure 1.3, raises several different questions. First of all: Is the sparsity level of a signal a sufficient characteristic for a unique solution? Moreover, what are the relations between n, m and the sparsity level s for this to hold and what are appropriate conditions to the measurement matrix Φ? At last, why and under which conditions does ℓ_1-minimization successfully or robustly recover the measured signal?

---

[2] A reconstruction was said to be *exact* if the ℓ_1-error between the true signal x and its reconstruction x° did not exceed 10^{-3}, i.e. if $\|x - x^\circ\|_1 \leqslant 10^{-3}$.

A general introduction to CS and its basic principles is given in Chapter 4 where there are also given answers to the above questions. In Chapter 7 the main results for the special application of CS in imaging mass spectrometry are presented. At last, in Chapter 8 the notion of CS appears in the context of CS based sporadic communication as the second application in this thesis.

1.4. Scientific contributions of the thesis

This thesis covers two different but parallel fields in both the pure and applied mathematics, namely *data compression* and *compressed sensing*.

Concerning data compression, commonly used techniques focusing on imaging mass spectrometry (IMS) data are discussed. Among these is peak picking which is used to detect only the most relevant peaks in the data spectra. The first scientific contribution of this work is done here as a peak-picking method (Chapter 6) is introduced which is much more sensitive in comparison to the standard approaches. Instead of considering the spectral point of view, it is changed to the spatial one and defines a measure to evaluate the structure of the mass images. Numerical results show the effectiveness of the proposed approach. The results were published in the following research article:

Alexandrov, T. and Bartels, A.: *Testing for presence of known and unknown molecules in imaging mass spectrometry.* Bioinformatics, 29(18):2335–2342, 2013.

The main part of this thesis and the second contribution is in the area of compressed sensing (CS). It is intended to give an introduction to the topic so that the basic principles as well as the results presented later for IMS are easily understandable. Regarding the latter, a novel CS model (Chapter 7) that allows the reconstruction of a full IMS dataset from only partial measurements is introduced. While reconstructing the data, its features are extracted in both the spectral and the spatial domain by peak-picking using the ℓ_1-norm as well as image denoising with the TV semi-norm, both of which are common IMS post-processing steps [9, 68].

A new robustness result that covers the robust recovery of both the spectra and the m/z-images in the CS setup is proven. Inspired by the application to IMS, the corresponding theorem combines for the first time stable ℓ_1- and TV-reconstruction results [47, 151]. For this, two different sparsity aspects are considered, namely in spectral and in spatial domain. This again differs from usual settings where only one sparsity aspect for the whole dataset is considered [99, 100, 150].

Numerical results on two different IMS datasets, namely a rat brain dataset and a rat kidney dataset, show the significance of the model as well as the robustness theorem. The results were published in

Bartels, A., Dülk, P., Trede, D., Alexandrov, T., and Maaß, P.: *Compressed sensing in imaging mass spectrometry.* Inverse Problems, 29(12):125015 (24pp), 2013.

and

Bartels, A., Trede, D., Alexandrov, T., and Maaß, P.: *Hybrid regularization and sparse reconstruction of imaging mass spectrometry data.* Proceedings of the 10th international conference on Sampling Theory and Applications, EURASIP, 189–192, 2013.

The third and last scientific contribution is a parameter choice rule for the elastic-net functional that is applied to sporadic communication (Chapter 8); a topic studied in communication engineering. As the word *sporadic* already indicates, sparsity plays a key role in the representative model. It describes the scenario of a number of nodes transmitting signals or messages only sporadically to a central aggregation point. Numerical results show that minimizing the elastic-net with the proposed parameter identification outperforms the often used OMP method for reconstructing the transmitted signals.

The parameter choice rule and the results presented in this thesis were developed in cooperation with the Department of Communications Engineering (ANT) of the Institute of Telecommunications and High-Frequency Technique (ITH) from the University of Bremen and published in

Schepker, H. F., Bockelmann, C., Dekorsy, A., Bartels, A., Trede, D., and Kazimierski, K. S.: *C-curve: A finite alphabet based parameter choice rule for elastic-net in sporadic communication.* IEEE Signal Processing Letters, 18(8):1443–1446, 2014.

All the methods and results described in this thesis were, unless otherwise stated, implemented in Matlab [140], version 8.1.0.604 (R2013a).

1.5. Organization of the thesis

The structure of the thesis is visualized in Figure 1.4 and described in the following: While Chapter 1 gave a general motivation for the analysis of compression techniques and compressed sensing, the following three chapters form the basis for the results of the later chapters.

Chapter 2 introduces the basic notation and results as well as the numerical approaches used within this thesis.

Chapters 3, 4 and 5 lay the foundation for the subsequent chapters by introducing compression as well as compressed sensing principles. In addition, a short introduction to mass spectrometry with the focus on MALDI-MS is given followed by a detailed explanation of the data sets used.

In Chapter 6, the compression techniques that were described in Chapter 3 are applied to mass spectrometry data. For this the spectral and spatial redundancies in IMS data are described. Besides well-known compression methods, a new peak-picking method is presented that is more sensitive compared to other approaches.

Chapter 7 contains the main results of the thesis. At first, a novel CS in IMS model is presented that takes into account the priors discussed in the previous Chapter 6. Second, a robustness result is given which combines the two so far separately studied ℓ_1- and TV-cases. At last, numerical results for different number of compressed measurements are shown for the rat brain dataset. The results for the rat kidney sample are left to the Appendix A.

Chapter 8 presents a known model for sporadic communication as well as a new parameter choice rule for the elastic-net functional, which is used for reconstructing the transmitted signals. Numerical results finally round off the chapter.

At last, Chapter 9 concludes with a brief summary and an outlook on possible future research.

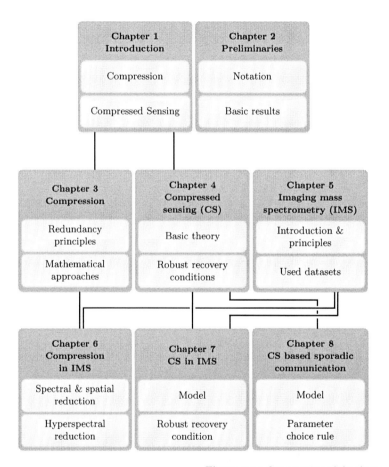

Figure 1.4.: Organization of the thesis.

2 | Preliminaries and concepts

This chapter introduces the basic notation as well as the basic principles concerning proximity operators and algorithms used in this thesis. The presentation of the latter relies on the textbooks [21, 178] and the articles [59, 60].

2.1. Notations

For $1 \leqslant p < \infty$, the matrix p-norm for a matrix $A = (A_{i,j}) \in \mathbb{R}^{m \times n}$ is denoted by

$$\|A\|_p^p = \sum_{i=1}^m \sum_{j=1}^n |A_{i,j}|^p$$

which is induced by the ℓ_p-vector norm

$$\|x\|_p^p = \|x\|_{\ell_p}^p = \sum_i |x_i|^p$$

for some $x \in \mathbb{R}^n$. For $p = \infty$ this norm is defined as $\|x\|_\infty = \max_i |x_i|$. For $p = 0$ this is

$$\|x\|_0 = \|x\|_{\ell_0} = |\text{supp}(x)| = |\{i \mid x_i \neq 0\}|.$$

Even though the latter is neither a norm nor a semi-norm it is referred to it as the ℓ_0-"norm" [71]. The corresponding ℓ_0-"norm" for matrices can be defined accordingly. In the case of $p = 2$, the matrix norm is the Frobenius norm, denoted by $\| \cdot \|_F$. This norm is generated by the inner product

$$\langle A, B \rangle = \text{trace}(AB^T) = \sum_{\substack{1 \leqslant i \leqslant m \\ 1 \leqslant j \leqslant n}} A_{i,j} B_{i,j}, \tag{2.1}$$

where $A = (A_{i,j}) \in \mathbb{R}^{m \times n}$ and $B = (B_{i,j}) \in \mathbb{R}^{m \times n}$. For a matrix $A = (A_{i,j,k}) \in \mathbb{R}^{m \times n \times r}$ this notation is canonically supplemented as

$$\|A\|_F^2 = \sum_{i=1}^{m} \sum_{j=1}^{n} \sum_{k=1}^{r} |A_{i,j,k}|^2.$$

The Kronecker product $\otimes$ of two matrices $A \in \mathbb{R}^{m \times n}$ and $B \in \mathbb{R}^{p \times q}$ is a $mp \times nq$ block matrix:

$$A \otimes B = \begin{pmatrix} A_{1,1}B & \dots & A_{1,n}B \\ \vdots & \ddots & \vdots \\ A_{m,1}B & \dots & A_{m,n}B \end{pmatrix}. \tag{2.2}$$

Lemma 2.1.1. *For a matrix $A \in \mathbb{R}^{m \times n}$ with s nonzero components it holds*

$$\|A\|_1 \leqslant \sqrt{s}\|A\|_F.$$

Proof. Let $E^{i,j} = (E_{k,l}^{i,j}) \in \mathbb{R}^{m \times n}$ be the matrix having zero entries except for the index (i,j) where it has a one if and only if $A_{i,j} \neq 0$. Since A has s nonzero components there are s matrices $E^{i,j}$ having a one at the corresponding tuple. With the expression $A = \sum_{i=1}^{m} \sum_{j=1}^{n} A_{i,j} E^{i,j}$ and the Cauchy-Schwarz inequality, it follows

$$\|A\|_1 \leqslant \sum_{i,j=1}^{m,n} |A_{i,j}| \sum_{k,l=1}^{m,n} |E_{k,l}^{i,j}|$$

$$\leqslant \sqrt{\sum_{i,j=1}^{m,n} |A_{i,j}|^2} \sqrt{\sum_{i,j=1}^{m,n} \left(\sum_{k,l=1}^{m,n} |E_{k,l}^{i,j}| \right)^2}$$

$$= \sqrt{s}\|A\|_F \qquad \qquad \square$$

The discrete total variation (TV) semi-norm of A is defined by

$$\|A\|_{TV} = \|\nabla A\|_1, \tag{2.3}$$

where $(\nabla A)_{i,j}$ denotes the discretized gradient. More precisely the discrete directional derivatives are given by

$$\begin{aligned} A_x &: \mathbb{R}^{m \times n} \to \mathbb{R}^{(m-1) \times n}, & (A_x)_{i,j} &= A_{i+1,j} - A_{i,j}, \\ A_y &: \mathbb{R}^{m \times n} \to \mathbb{R}^{m \times (n-1)}, & (A_y)_{i,j} &= A_{i,j+1} - A_{i,j}. \end{aligned} \tag{2.4}$$

The discrete gradient transform $\nabla : \mathbb{R}^{m \times n} \to \mathbb{R}^{m \times n \times 2}$ is then defined component wise as follows

$$(\nabla A)_{i,j} = \begin{cases} ((A_x)_{i,j}, (A_y)_{i,j}), & 1 \leqslant i \leqslant m-1, \ 1 \leqslant j \leqslant n-1 \\ (0, (A_y)_{i,j}), & i = m, \ 1 \leqslant j \leqslant n-1 \\ ((A_x)_{i,j}, 0), & 1 \leqslant i \leqslant m-1, \ j = n \\ (0,0), & i = m, \ j = n. \end{cases}$$

Remark 2.1.2. The TV norm is a semi-norm since the null-space of the mapping $A \mapsto \|A\|_{TV}$ consists of all constant matrices A and not only of the zero matrix, as it is required in the definition of the norm.

In this thesis, the anisotropic variant of the total variation norm will be used which is given by

$$\|\nabla A\|_1 = \sum_{i,j} \|(\nabla A)_{i,j}\|_1. \tag{2.5}$$

In the isotropic case one would have

$$\|\nabla A\|_2 = \sum_{i,j} \|(\nabla A)_{i,j}\|_2,$$

which is equivalent to the anisotropic case (2.5) up to a factor $\sqrt{2}$. The results presented in this thesis are given for the anisotropic case only and the corresponding analysis for the isotropic case is left open.

The notation $x \lesssim y$ means that there exists some constant $C > 0$ such that $x \leqslant Cy$. The expression $x \gtrsim y$ is defined accordingly. In addition, the notation $\mathbb{R}_+ = \{x \in \mathbb{R} \mid x \geqslant 0\}$ is used.

Lemma 2.1.3 (Inequality of arithmetic and geometric means). *For the nonnegative real numbers $x_1, x_2, \ldots, x_n$ there holds*

$$\sqrt[n]{x_1 \cdot x_2 \cdot \ldots \cdot x_n} \leqslant \frac{x_1 + x_2 + \ldots + x_n}{n}.$$

Equality holds if and only if $x_1 = x_2 = \ldots = x_n$.

Proof. See [194] for a proof via induction. $\square$

2.2. Proximity operators and algorithms

In this section, basic principles about proximity operators as well as a brief presentation of the relevant algorithms are presented. They are mainly used in Chapter 7 where a compressed sensing model for the application in imaging mass spectrometry is minimized.

Let $\mathbb{R}_\infty = \mathbb{R} \cup \{\infty\}$ and X be a normed space. $f : X \to \mathbb{R}_\infty$ is called *proper* if $\mathrm{dom}(f) := \{x \in X \mid f(x) < \infty\} \neq \varnothing$. f is *lower semicontinuous* if $\liminf_{x \to x_0} f(x) \geq f(x_0)$. It is clear that every continuous function is also lower semicontinuous.

Example 2.2.1. A simple example for a lower semicontinuous function is given by

$$f : \mathbb{R} \longrightarrow \mathbb{R}, \quad f(x) = \begin{cases} 0, & x \leq 0 \\ 1, & x > 0. \end{cases}$$

This example can also easily change to a counterexample by modifying the function f to be 0 if $x < 0$ and to be 1 if $x \geq 0$.

It is often difficult to prove lower semicontinuity with the use of its definition. Therefore, the following lemma gives a useful result.

Lemma 2.2.2 ([21, Lemma 1.24]). *Let X be a normed space and let $f : X \to [-\infty, \infty]$. Then the following are equivalent*

(i) f is lower semicontinuous.

(ii) For every $\xi \in \mathbb{R}$ the (level) set

$$\mathrm{lev}_{\leq \xi}(f) := \{x \in X \mid f(x) \leq \xi\} \tag{2.6}$$

of f at height $\xi \in \mathbb{R}$ is closed in X.

A set $\mathcal{C}$ in the normed space X is said to be *convex* if for all $x, y \in \mathcal{C}$ and $\lambda \in (0, 1)$ the convex combination $(1 - \lambda)x + \lambda y$ is still an element of $\mathcal{C}$. A function f is said to be *convex* on a convex set $\mathcal{C}$ if it holds

$$f((1 - \lambda)x + \lambda y) \leq (1 - \lambda)f(x) + \lambda f(y) \tag{2.7}$$

for distinct points $x, y \in C$ and $\lambda \in (0,1)$. f is called *strictly convex* if the right hand side in (2.7) is strictly greater for all $x, y \in C$ and $\lambda \in (0,1)$.

In the following let $\mathcal{H}$ be a Hilbert space. The notation $\Gamma(\mathcal{H})$ denotes the set of all lower semicontinuous convex functions from $\mathcal{H}$ to $\mathbb{R}_\infty$. In addition $\Gamma_0(\mathcal{H})$ is a subset of $\Gamma(\mathcal{H})$ where the functions are also proper. The *(Fenchel) conjugate* of a function $f \in \Gamma_0(\mathcal{H})$ is defined by $f^* := \sup_{x \in \mathrm{dom}(f)}(\langle x, \cdot \rangle - f(x))$. The *subdifferential* of f at position $x \in \mathrm{dom}(f)$ is the set

$$\partial f(x) := \big\{ y \in \mathcal{H} \mid \forall z \in \mathcal{H} : \langle z - x, y \rangle + f(x) \leqslant f(z) \big\}.$$

The elements lying in $\partial f(x)$ are called *subgradients*. If f is differentiable at x with gradient $\nabla f(x)$, then $\partial f(x) = \{\nabla f(x)\}$.

Theorem 2.2.3 (Optimality condition [36, Theorem 6.43]). *Let $f : \mathcal{H} \to \mathbb{R}_\infty$ be a convex functional. Then it holds*

$$p^\diamond := \operatorname*{argmin}_{p \in \mathcal{H}} f(p) \quad \Longleftrightarrow \quad 0 \in \partial f(p^\diamond).$$

The next lemma states under which condition a composition of a function $f \in \Gamma_0(\mathcal{H})$ with a bounded affine operator $A = \Phi \cdot - y$ is an element in $\Gamma_0(\mathcal{H})$. For this let, for a function $f : \mathcal{H} \to \mathbb{R}_\infty$, $\mathrm{Im}(f) = \{z \in \mathbb{R}_\infty \mid \exists x \in X : f(x) = z\}$ be the image of f. The interior of a subset $D \subseteq \mathcal{H}$ is denoted by $\mathrm{int}(D)$ and describes the largest open set contained in D.

Lemma 2.2.4 ([88, Lemma 2]). *Let $f \in \Gamma_0(\mathcal{H})$. If Φ is a bounded linear operator such that $\mathrm{int}(\mathrm{dom}(f) \cap \mathrm{Im}(A)) \neq \varnothing$, then $f \circ A \in \Gamma_0(\mathcal{H})$.*

Now let $f \in \Gamma_0(\mathcal{H})$. Then, for every $x \in \mathcal{H}$ the *proximity operator* [145] is defined as the operator $\mathrm{prox}_f : \mathcal{H} \to \mathcal{H}$ for which $\mathrm{prox}_f(x)$ is the unique point in $\mathcal{H}$ that satisfies

$$\mathrm{prox}_f(x) := \operatorname*{argmin}_{y \in \mathcal{H}} \frac{1}{2}\|x - y\|_2^2 + f(y).$$

The existence of a minimizer of the function f is guaranteed because f is convex and lower semicontinuous. The uniqueness of $\mathrm{prox}_f(x)$ follows from the additional quadratic data fidelity term which makes the underlying functional strictly convex.

In Chapter 7, where a compressed sensing model for the application in imaging mass spectrometry is studied, the *parallel proximal splitting algorithm* (PPXA) will be used to solve an optimization problem in form of a weighted sum with weights $\gamma_i > 0$, $i = 1, \ldots, t$,

$$\operatorname*{argmin}_{x \in \mathcal{H}} \sum_{i=1}^{t} \gamma_i f_i(x). \tag{2.8}$$

PPXA is an iterative method for minimizing a finite sum of lower semi-continuous convex functions [59]. It is easy to implement and has the possibility to be parallelized. At each iteration of the algorithm one needs to calculate the proximity operator of each function and to average their results for updating the previous iterate. It has been shown in [59] that under some assumptions every sequence $(x_n)_{n \in \mathbb{N}}$ generated by the PPXA converges weakly to a solution of (2.8).

Algorithm 1: Parallel Proximal Splitting Algorithm for solving (2.8)

Input: $\eta \in \,]0,1[\,, \gamma_j > 0, (\omega_j)_{1 \leqslant j \leqslant t} \in \,]0,1[^t$ s.t. $\sum_{j=1}^{t} \omega_j = 1$

Initializations: $k = 0$; $y_{1,0}, \ldots, y_{t,0} \in \mathbb{R}^n$, $x_0 = \sum_{j=1}^{t} \omega_j y_{j,0}$

repeat

    for $j = 1 : t$ **do**

        $p_{j,k} = \operatorname{prox}_{\gamma_j f_j / \omega_j}(y_{j,k})$

    end

    $q_k = \sum_{j=1}^{t} \omega_j p_{j,k}$

    $\eta \leqslant \lambda_k \leqslant 2 - \eta$

    for $j = 1 : t$ **do**

        $y_{j,k+1} = y_{j,k} + \lambda_k(2q_k - x_k - p_{j,k})$

    end

    $x_{k+1} = x_k + \lambda_k(q_k - x_k)$

until *convergence*

In this work, a weighted sum of four different functions f_j, $j = 1, \ldots, 4$, in the case of $\mathcal{H} = \mathbb{R}^n$ (or analogously $\mathcal{H} = \mathbb{R}^{m \times n}$) occur for which the expressions

$$p_j = \operatorname{prox}_{\gamma_j f_j}(y) = \operatorname*{argmin}_{p \in \mathbb{R}^n} \frac{1}{2}\|y - p\|_2^2 + \gamma_j f_j(p) \tag{2.9}$$

in the PPXA need to be calculated. These functions are

(i) the ℓ_1-norm, i.c. $f_1 = \| \cdot \|_1$,

(ii) the TV-norm or a sum of those, i.e. $f_2 = \| \cdot \|_{TV}$,

(iii) the projection onto the positive orthant, i.e. $f_3 = \iota_{\mathbb{R}^n_+}$, where $\iota_{\mathcal{C}}$ denotes the indicator function with respect to the set $\mathcal{C}$

$$\iota_{\mathcal{C}}(y) = \begin{cases} 0 & \text{if } y \in \mathcal{C} \\ +\infty & \text{otherwise} \end{cases}, \tag{2.10}$$

(iv) the projection onto the closed ℓ_2-ball of radius ε with respect to a bounded affine operator

$$A : \mathbb{R}^n \to \mathbb{R}^m, \quad Az = \Phi z - w, \tag{2.11}$$

i.e. $f_4 = \iota_{B_2^\varepsilon} \circ A$. There, $\Phi \in \mathbb{R}^{m \times n}$ is a linear operator with columns $\varphi_i \in \mathbb{R}^m$, $i = 1, \ldots, n$, and fulfills

$$c_1 \|z\|_2^2 \leqslant \|\Phi z\|_2^2 \leqslant c_2 \|z\|_2^2 \tag{2.12}$$

with real-valued constants $c_1 \leqslant c_2 < \infty$.

Remark 2.2.5. In case there exist constants c_1, c_2 such that the chain of inequalities (2.12) holds for $0 < c_1 \leqslant c_2 < \infty$, then Φ is called a *frame*. A frame is basically a generalization of a basis and allows linearly dependencies within the set of the underlying vectors. In addition, it is still a generating system for the given space. In case of Φ in (2.12) and $m < n$ this means that the n columns $\varphi_i \in \mathbb{R}^m$ are linearly dependent, but still build a spanning set of $\mathbb{R}^m$.

In the next, it will be argued about each function f_i that it is a proper lower semicontinuous convex function, meaning that $f_i \in \Gamma_0(\mathbb{R}^n)$. More-over, it is discussed how the corresponding proximity operators (2.9) can be calculated.

In the case of $f_1 = \| \cdot \|_1$ first note that every norm $\| \cdot \| : \mathbb{R}^n \to \mathbb{R}_+$ is a convex and continuous function due to the triangle inequality. Every norm is also obviously proper. Therefore it is clear that $f_1 = \|\cdot\|_1$ is a function in $\Gamma_0(\mathbb{R}^n)$. Then, for a given $y \in \mathbb{R}^n$ and coefficient γ_1 define

$$p_1 := \operatorname{prox}_{\gamma_1 \|\cdot\|_1}(y) = \operatorname*{argmin}_{p \in \mathbb{R}^n} \frac{1}{2}\|y - p\|_2^2 + \gamma_1 \|p\|_1.$$

Since f_1 is not differentiable in $p = 0$, one needs to make use of the subdifferential. The optimality condition from Theorem 2.2.3 for p_1 then reads

$$0 \in \nabla\left(\frac{1}{2}\|y - p_1\|_2^2\right) + \partial(\gamma_1 \|p_1\|_1) = p_1 - y + \gamma \partial \|p_1\|_1.$$

For simplicity, the following consideration is made only with respect to the i-th component $p_{1,i}$ of p_1. In case of $p_{1,i} \neq 0$ the subdifferential $\partial \|p_{1,i}\|_1$ is single-valued, $\partial \|p_{1,i}\|_1 = \operatorname{sign}(p_{1,i})$, where $\operatorname{sign}(\cdot)$ describes the signum function that is defined for real-valued $y_i \in \mathbb{R}$ as

$$\operatorname{sign}(y_i) = \begin{cases} 1 & \text{if } y_i > 0 \\ 0 & \text{if } y_i = 0 \\ -1 & \text{if } y_i < 0 \end{cases} \tag{2.13}$$

This leads to $0 = p_{1,i} - y_i + \gamma_1 \operatorname{sign}(p_{1,i})$, or

$$p_{1,i} = y_i - \gamma_1 \operatorname{sign}(p_{1,i}). \tag{2.14}$$

If $p_{1,i} < 0$, then $y_i < -\gamma_1$. Else, if $p_{1,i} > 0$, then $y_i > \gamma_1$ and thus $|y_i| > \gamma_1$. As $\gamma_1 > 0$, it follows $\operatorname{sign}(p_{1,i}) = \operatorname{sign}(y_i)$ and (2.14) becomes

$$p_{1,i} = y_i - \gamma_1 \operatorname{sign}(y_i),$$

In the other case $p_{1,i} = 0$, it is $0 \in -y_i + \gamma_1[-1, 1]$, i.e. $|y_i| \leqslant \gamma_1$. Putting this all together leads to

$$p_{1,i} = \operatorname{prox}_{\gamma_1 \|\cdot\|_1}(y_i) = \begin{cases} 0 & \text{if } |y_i| \leqslant \gamma_1 \\ y_i - \gamma_1 \operatorname{sign}(y_i) & \text{if } |y_i| > \gamma_1. \end{cases}$$

This is also known as the soft thresholding [21] with threshold γ_1 and can be rewritten as

$$\begin{aligned} p_1 = \operatorname{prox}_{\gamma_1 \|\cdot\|_1}(y) &= \left(\max\left\{0, \left(1 - \frac{\gamma_1}{|y_i|}\right)\right\} y_i\right)_{1 \leqslant i \leqslant n} \\ &= \left(\max\left\{0, |y_i| - \gamma_1\right\} \operatorname{sign}(y_i)\right)_{1 \leqslant i \leqslant n}. \end{aligned} \tag{2.15}$$

For the proximity operator of the TV semi-norm, i.e. $f_2 = \|\cdot\|_{TV}$, first note that $f_2 \in \Gamma_0(\mathbb{R}^n)$ [15]. Unfortunately, there is no direct closed expression available [59]. Therefore, one needs to approximate the solution to (2.9). For this a fast gradient projection method from [23] is used, which is an accelerated variant of Chambolle's algorithm [53].

For the proximity operator of the indicator function ι_C (2.10) for a set $C \subseteq X$ it needs to be explained that $\iota_C \in \Gamma_0(\mathbb{R}^n)$. For the lower semicontinuity choose a fixed but arbitrary $\xi \in \mathbb{R}$. Then, by definition, the level set (2.6)

$$\text{lev}_{\leqslant \xi}(\iota_C) = \{x \in X \mid \iota_C(x) \leqslant \xi\}$$

is empty if $\xi < 0$. Otherwise, i.e. if $\xi \geqslant 0$, it is $\text{lev}_{\leqslant \xi}(\iota_C) = C$. According to Lemma 2.2.2, the indicator function ι_C is then lower semicontinuous if and only if C is closed in X. In addition, if C is convex, it directly follows that ι_C is a convex function. If C is non-empty, ι_C is also proper. Here $C = \mathbb{R}_+^n$ is a non-empty closed convex set and it follows $\iota_C \in \Gamma_0(\mathbb{R}^n)$. The proximal operator of the indicator function $f_3 = \iota_{\mathbb{R}_+^n}$ is unique and given as the orthogonal projection onto $\mathbb{R}_+^n$:

$$\text{prox}_{\gamma_3 \iota_{\mathbb{R}_+^n}}(y) = \left(\max\{0, y_i\}\right)_{1 \leqslant i \leqslant n}.$$

For the last function f_4 one first argues on a similar way as for f_3. Since the ℓ_2-ball of radius ε is a closed convex set, it follows with Lemma 2.2.2 that $\iota_{\mathcal{B}_2^\varepsilon} \in \Gamma_0(\mathcal{B}_2^\varepsilon)$. Since the operator $A : \mathbb{R}^n \to \mathbb{R}^m$ in (2.11) is assumed to be a bounded linear operator, it follows together with Lemma 2.2.4 that $f_4 = \iota_{\mathcal{B}_2^\varepsilon} \circ A \in \Gamma_0(\mathbb{R}^n)$. For the proximity operator of $\gamma_4 f_4$ with $f_4 = \iota_{\mathcal{B}_2^\varepsilon} \circ A$ one needs to solve

$$\text{prox}_{\gamma_4(\iota_{\mathcal{B}_2^\varepsilon} \circ A)}(y) = \underset{p \in \mathbb{R}^n}{\text{argmin}} \, \frac{1}{2}\|y - p\|_2^2 + \gamma_4(\iota_{\mathcal{B}_2^\varepsilon} \circ A)(p). \qquad (2.16)$$

The calculation of (2.16) is then done by the following algorithm.

The proximity operator (2.16) is then given as $y - \Phi^T u_{n_{\text{iter}}}$. In addition, this iteration converges linearly to $\text{prox}_{\gamma_4(\iota_{\mathcal{B}_2^\varepsilon} \circ A)}(y)$ [88]. Algorithm 2 requires calculating two times the proximity operator of the form

$$\text{prox}_{\gamma \iota_{\mathcal{B}_2^\varepsilon}}(y) = \underset{p \in \mathbb{R}^n}{\text{argmin}} \, \frac{1}{2}\|y - p\|_2^2 + \gamma \iota_{\mathcal{B}_2^\varepsilon}(p),$$

23

Algorithm 2: Proximal operator of precomposition with an affine
operator [Algorithm 21, [178]]

Initializations: c_1, c_2 via (2.12),
$$u_0 \in \mathrm{dom}(\iota_{\mathcal{B}_2^\varepsilon}^*), \mu \in (0, 2/c_2), \theta_0 = 0, \xi_0 = 0$$

for $i = 0, \ldots, n_{\text{iter}} - 1$ **do**

    Set $\rho_i = \mu(1 + c_1\theta_i)$

    $\eta_i = \theta_i\left(I - \mathrm{prox}_{\gamma_4\theta_i^{-1}\iota_{\mathcal{B}_2^\varepsilon}}\right)\left(\frac{u_0 - \xi_i}{\theta_i}\right)$

    Set $a_i = \frac{\rho_i + \sqrt{\rho_i^2 + 4\rho_i\theta_i}}{2}$ and $\omega_i = \frac{\theta_i u_i + a_i \eta_i}{\theta_i + a_i}$

    $u_{i+1} = \frac{\mu}{2}(I - \mathrm{prox}_{2\gamma_4\mu^{-1}\iota_{\mathcal{B}_2^\varepsilon}})(2\mu^{-1}\omega_i + (\Phi(y - \Phi^*\omega_i) - w))$

    $\xi_{i+1} = \xi_i - a_i(\Phi(y - \Phi^*u_i - w))$

    $\theta_{i+1} = \theta_i + a_i$

end

---

whose solution is simply given as the orthogonal projection onto $\iota_{\mathcal{B}_2^\varepsilon}$,
namely

$$\mathcal{P}_{\iota_{\mathcal{B}_2^\varepsilon}}(y) = \begin{cases} y & \text{if } \|y\|_2 \leqslant \varepsilon \\ \varepsilon\frac{y}{\|y\|_2} & \text{otherwise} \end{cases}.$$

3 | Data compression

This chapter summarizes the key aspects of data compression, as they occur in the present work. It covers an explanation and a classification of compression and also discusses relevant compression techniques from the mathematical point of view. The presentation, especially that for the first two topics, mainly follows that from the monographs [159, 164] and [178].

3.1. What is compression?

As already motivated in Section 1.2, the interest on data compression has grown over the last ten years as more and more data is collected. But what exactly does *compression* mean in general and in the context of this thesis? As written by Sayood in [166], "data compression is the art or science of representing information in a compact form" by "identifying and using structures that exist in the data". Methods may have different approaches for extracting only the information that is relevant and removing everything else, but they all have in common that they treat *redundancies* in the data [164]. And this is how compression is generally understood: As a way to get rid of redundancy and keep the rest. As it is well explained in [159], redundancies in data such as in digital images can generally be subdivided in the following three classes:

1. *Spatial redundancy*: Due to correlation between neighboring pixels.

2. *Spectral redundancy*: Due to correlation between different color planes (e.g. in an RGB color image) or spectral bands (e.g. aerial photographs in remote sensing).

3. *Temporal redundancy*: Due to correlation between different frames in a sequence of images.

This thesis is only concerned with the first two points. In Chapter 6, compression techniques are presented which essentially focus on spatial redundancy of the mass images (Section 6.3) and spectral redundancy of the mass spectra (Section 6.2). Under the same aspects, a CS approach for IMS data is analyzed in Chapter 7.

Removing redundancies in given data is connected with the task of deciding whether the possibility for a perfect reconstruction is requested or not. If the answer is yes, then *lossless compression* techniques should be used. Otherwise, *lossy compression* can be used and usually implies better compression rates (cf. (3.1)) than in the lossless case. For an overview for both types of algorithms, refer to the monographs [159] and [164] and the references therein.

At the end of this section it should be remarked that whenever information is important or not, one is usually faced with the full original data and then has to extract only what is relevant. In contrast, the compression step in CS has already taken place during the acquisition process. While compression here means, again, a reduction of the data (as one measures less than the given data dimension), redundancies are treated during the measurement process, rather than during the subsequent minimization.

3.2. Compression and quality measures

Several quantities exist to measure the *performance* of data compression. One common quantity is the *compression ratio* or *compression rate* that is defined by

$$\text{Compression ratio} = \frac{\text{size of output}}{\text{size of input}}. \tag{3.1}$$

It is clear that the smaller the size of the output, the smaller the compression ratio (3.1) and the *stronger* is the compression effect. As an example, a value of 0.2 means that only 20% of the original data size is left after compression. As this ratio only evaluates the performance of the reduction, measuring the *quality* of the reduced data in comparison with the original data is left open. Here, two commonly used dimensionless error metrics in image compression are presented. One is the *peak signal to noise ratio* (PSNR) that is defined over the *mean square error* (MSE), which is given as

$$\text{MSE} = \frac{1}{mn} \sum_{i=1}^{m} \sum_{j=1}^{n} (X_{i,j} - \tilde{X}_{i,j})^2,$$

where $X \in \mathbb{R}^{m \times n}$ and $\tilde{X} \in \mathbb{R}^{m \times n}$ denote the discrete input image and the output, or reconstructed image, respectively. If the elements of the image X are normalized to the range $[0, B]$, the PSNR is defined as

$$\text{PSNR} = 20 \log_{10} \frac{B}{\sqrt{\text{MSE}}}, \tag{3.2}$$

and often expressed in *decibels* (dB). If the difference between the original and the reconstructed image $X - \tilde{X}$ is small, the PSNR is large. PSNR values are always mentioned if an algorithm is tested for several compression ratios or if one compares several algorithms with each other. Note that the value for one single compression step only is meaningless. Typical values for the PSNR ranges between 20 and 40 [164].

Even though the PSNR (3.2) is very simple to calculate, it is described in [164] as only having "a limited, approximate relationship with the perceived errors noticed by the human visual system" and therefore "do not provide a guarantee that viewers will like the reconstructed image". This observation was mathematically analyzed in [200, 201], where the authors developed a new measure that is better adapted to the human visual system called the *structural similarity* (SSIM) index. For two different images $X \in \mathbb{R}^{m \times n}$ and $Y \in \mathbb{R}^{m \times n}$ it is defined as

$$\text{SSIM}(X, Y) = \frac{(2\mu_X \mu_Y + c_1)(2\sigma_{XY} + c_2)}{(\mu_X^2 \mu_Y^2 + c_1)(\sigma_X^2 + \sigma_Y^2 + c_2)}, \tag{3.3}$$

where μ_X and μ_Y are the mean intensities of all pixels in the images X and Y. The notations σ_X and σ_Y denote their standard deviations and are, on the example of the matrix X, estimated by

$$\sigma_X = \left(\frac{1}{mn-1} \sum_{i=1}^{m} \sum_{j=1}^{n} (X_{i,j} - \mu_X)^2 \right).$$

The term σ_{XY} describes the covariance of X and Y which is estimated by

$$\sigma_{XY} = \frac{1}{mn-1} \sum_{i=1}^{m} \sum_{j=1}^{n} (X_{i,j} - \mu_X)(Y_{i,j} - \mu_Y).$$

The variables c_1, c_2 are given to stabilize the division. The SSIM index (3.3) is bounded by $-1 < \text{SSIM}(X,Y) \leqslant 1$ and takes its maximum $\text{SSIM}(X,Y)$ if and only if $X = Y$ [200]. This means that the closer the index to 1, the better the reconstruction according to the SSIM measure.

3.3. Mathematical techniques

3.3.1. ℓ_0 and ℓ_1 minimization

Finding redundancies in data for compression goes hand in hand with extracting the most relevant features, as described in Section 3.1. It needs to mentioned that here and throughout the thesis the focus will be on *sparsity* related compression only, meaning that one searches for only few non-zero coefficients with respect to an representative system. In fact, usual compression techniques comprise of more working steps or *coding schemes* [159, 164, 184].

Nevertheless, for the purpose of this work the following notion of a *dictionary* will be used

Definition 3.3.1. Let $x \in \mathbb{R}^n$. A *dictionary* is a matrix $D \in \mathbb{R}^{n \times p}$ containing p column vectors $d_i \in \mathbb{R}^n$, $i = 1, \ldots, p$, such that there exists a coefficient vector $\alpha \in \mathbb{R}^p$ such that $x = \sum_i \alpha_i d_i$. The elements d_i are also called *atoms*. If $p > n$, D is called an *overcomplete dictionary*.

The columns in D are allowed to be linearly dependent. Hence, a dictionary does not need to be a basis.

With this term in hand, reconstructing the most relevant features can be mathematically translated to the problem of giving a signal $x \in \mathbb{R}^n$ (e.g. a reshaped image) and seeking for a sparse coefficient vector $\lambda \in \mathbb{R}^p$ with respect to a given (possible overcomplete) dictionary $\Psi \in \mathbb{R}^{n \times p}$ with $p \geqslant n$, such that

$$x = \Psi\lambda. \tag{3.4}$$

In case of an orthogonal matrix $\Psi \in \mathbb{R}^{n \times n}$, the solution λ is unique. If $p > n$, the system (3.4) becomes under-determined and additional information on the solution λ is required. The dictionary Ψ is column wise composed of atoms ψ_i, $i = 1, \ldots, p$, and is set a priori with the goal that the signal x should have a good representation with only a few taken atoms, i.e. $\|\lambda\|_0 \ll n < p$, leading to compression.

Remark 3.3.2. It needs to be mentioned that their is a trade off between adding extra columns in Ψ and computation time since the unknown coefficient vector will increase in size as well. Moreover, some dictionaries might have computational advantages while others might be more suitable for extracting the main details in the data. Examples for possible choices for the atoms of Ψ are Fourier-basis elements, wavelets [65, 136], curvelets [177], and the shearlets [119], to mention only a few.

As mentioned in Section 1.2, the JPEG 2000 format is an example for both lossy and lossless compression and applies the wavelet transform within all its different compression steps [184]. An application of an only sparsity related wavelet based compression is given in Figure 3.1, where the right image is a highly compressed version of the left original image of size 1024×1024, as only the largest 7 % of all wavelet coefficients are taken into account. The PSNR (3.2) and SSIM (3.3) result in 37.04 dB and 0.9993, respectively. The reason why only a few coefficients are necessary lies in the general observation that photographs especially are often composed of homogenous areas while finer details make up only a small part of the photo. With regard to the bird example in Figure 3.1 it is observable that the body of the bird and the background are quite homogenous and fill most part of the photo. Details, such as in the face and feather of the bird, make up a very small portion of what is actually visible. Consequently, an image is already represented well by only a very few such constant regions and a small amount of finer information.

If one knows that λ is sparse, i.e. that if most of its entries are equal to zero, then a first suggestion would potentially be to solve the following ℓ_0 minimization problem

$$\underset{\tilde{\lambda} \in \mathbb{R}^p}{\operatorname{argmin}} \|\tilde{\lambda}\|_0 \text{ subject to } \Psi \tilde{\lambda} = x. \tag{3.5}$$

Figure 3.1.: An example for wavelet compression. The left image of size 1024×1024 shows a photograph of a european robin and the right one is a compressed version of it. Only 7 % of all wavelet coefficients are used for reconstruction.

Another possible different but similar problem formulation would take into account the knowledge that λ is s-sparse

$$\operatorname*{argmin}_{\tilde{\lambda} \in \mathbb{R}^p} \| \Psi \tilde{\lambda} - x \|_2^2 \text{ subject to } \| \tilde{\lambda} \|_0 \leqslant s. \tag{3.6}$$

Unfortunately, solving these non convex problems with incorporated ℓ_0-"norm" has proven to be NP-hard [138, 148]. Nevertheless, there are methods, such as greedy or thresholding-based algorithms, that still aim to find a solution to (3.5) or variants of it. A prominent greedy method is the *orthogonal matching pursuit* (OMP) [137, 155], which iteratively updates a target support and the solution $\tilde{\lambda}$ such that y is well represented via a linear combination of only few basis vectors in Ψ. The size of the support depends on the (expected) sparsity level s of the given signal x. Hence, OMP can be seen as an approximation to (3.6). One of the thresholding-based methods is the *iterative hard thresholding* (IHT) [28, 29]. IHT basically starts from the equation (3.4) which can be rewritten as $\Psi^T \Psi \lambda = \Psi^T x$. The latter then becomes the following fixed-point equation

$$\lambda = (I - \Psi^T \Psi)\lambda + \Psi^T x = \lambda + \Psi^T (x - \Psi \lambda).$$

As one aims in taking only the s largest absolute entries, one simply eliminates all other smaller entries by setting them to zero via the nonlinear thresholding operator H_s,

$$\lambda = H_s(\lambda + \Psi^T(x - \Psi\lambda)). \tag{3.7}$$

Remark 3.3.3. It was shown in [28] that the fixed point iteration (3.7) converges to a local minimum of problem (3.6) provided that $\|\Psi\|_2 < 1$.

Note that both the OMP and the IHT are not a contradiction to the mentioned NP-hardness of problem (3.5), because they are only heuristics in general and solve (3.5) only for certain instances.

Instead of one of the ℓ_0 formulations, the following convex relaxation can be considered, which is also known as the *basis pursuit* (BP) [56],

$$\underset{\tilde{\lambda} \in \mathbb{R}^p}{\operatorname{argmin}} \|\tilde{\lambda}\|_1 \quad \text{subject to} \quad \Psi\tilde{\lambda} = x. \tag{3.8}$$

Due to its convexity, it has the advantage that it can be solved by various convex optimization techniques [22, 24, 27–29, 40, 59, 149]. Moreover, the ℓ_1 formulation still promotes a sparse solution due to the geometry of the ℓ_1-ball, see Figure 3.2. There it is also motivated that an ℓ_2-variant of (3.5) and (3.8) respectively tends to non-sparse solutions.

If the signal x is corrupted with some bounded additive noise n, i.e. $x = \Psi\lambda + n$, $\|n\|_2 \leq \gamma$, then one usually tries to solve either the convex optimization problem

$$\underset{\tilde{\lambda} \in \mathbb{R}^p}{\operatorname{argmin}} \|\tilde{\lambda}\|_1 \quad \text{subject to} \quad \|\Psi\tilde{\lambda} - x\|_2 \leq \gamma. \tag{3.9}$$

or the following closely related called *basis pursuit denoising* (BPDN) [56] convex minimization problem

$$\underset{\tilde{\lambda} \in \mathbb{R}^p}{\operatorname{argmin}} \frac{1}{2}\|\Psi\tilde{\lambda} - x\|_2^2 + \alpha\|\tilde{\lambda}\|_1, \tag{3.10}$$

where $\alpha \geq 0$ is the regularization parameter. In literature, BPDN is additionally known as the *least absolute shrinkage and selection operator* (LASSO) [186] and as the ℓ_1-regularized *Tikhonov* [33–35, 66, 126, 188] variant.

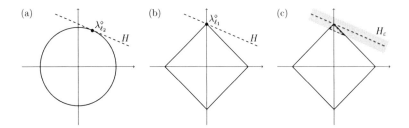

Figure 3.2.: Illustration of the ℓ_1-recovery. (a) Solving the optimization problem with respect to the ℓ_2-norm, where H denotes the set of all λ with the same value x under the multiplication with Ψ, $H = \{\lambda \mid \Psi\lambda = x\}$. Due to the geometry of the ℓ_2-ball, the solution $\lambda^\diamond_{\ell_2}$ tends to be non-sparse. (b) In contrast, the ℓ_1-ball promotes sparsity of the ℓ_1 solution $\lambda^\diamond_{\ell_1}$. (c) In the present of noise, a feasible solution lies in the intersection of $H_\varepsilon = \{\lambda \mid \|\Psi\lambda - x\|_2 \leqslant \varepsilon\}$ and the ℓ_1-ball.

Remark 3.3.4. The general relation of both formulations (3.9) and (3.10) is treated in Proposition 3.2 and Theorem B.28 in [93]. Let $\tilde{\lambda}_\alpha$ be the solution to (3.10) for some $\alpha > 0$. If one chooses the regularization parameter α according to the Morozov's discrepancy principle [32, 146], i.e. such that $\|\Psi\tilde{\lambda}_\alpha - x\| = \gamma$, then (3.9) and (3.10) are equivalent.

3.3.2. TV minimization

Lastly in this section, a very prominent variant of (3.10) for piecewise smooth images should be mentioned. For this let $X \in \mathbb{R}^{n_x \times n_y}$ be an image. Ψ is the identity and it is a priori known that X is piecewise smooth and therefore fulfills the spatial redundancy principle from Section 3.1. This can be understand in such way that the image gradient ∇X is sparse, leading to the *total variation* [1] norm $\|X\|_{TV} = \|\nabla X\|_1$ (cf. Chapter 2) and, with $\alpha > 0$, to the minimization problem

$$\operatorname*{argmin}_{\tilde{X} \in \mathbb{R}^{n_x \times n_y}} \frac{1}{2}\|\tilde{X} - X\|_F^2 + \alpha\|\tilde{X}\|_{TV}. \tag{3.11}$$

This problem is literally known as the Rudin, Osher and Fatemi (ROF) model and was introduced in [163]. A solution to (3.11) is a smoothed version of the probably noisy image X. In addition, the image edges (discontinuities) are supposed to be preserved under this minimization.

Remark 3.3.5. It needs to be mentioned that in the given discrete setting, differentiating the just mentioned discontinuities is possible as one calculates the gradient by finite differences. In finite dimensions, however, the expression $\|\nabla X\|_1$ typically reads $\int_\Omega |DX|$ and needs to be understood in the distributive sense where Ω is a bounded open subset of $\mathbb{R}^2$ and D is a gradient in distributional sense [15].

The regularization parameter $\alpha > 0$ in (3.11) determines the effect of smoothing. By increasing α, noise is removed and contrast is lost as well. In addition, it is probably the case that small structures in the image disappear which is then a consequence of the TV smoothing effect. For more information on total variation problems, refer to the articles [53, 163, 199] as well as the review chapter in [38] and the references therein.

3.3.3. Nonnegative matrix factorization

This section gives a short motivation for (nonnegative) matrix factorization. The purpose is to present a tool for the compressed representation of large datasets such as hyperspectral data in imaging mass spectrometry, cf. Chapter 6.

Let $X \in \mathbb{R}^{n \times m}$ be a known matrix and additionally $P \in \mathbb{R}^{n \times \rho}$ and $S \in \mathbb{R}^{\rho \times m}$ unknown matrices. Then the matrix Y can be expressed as the *matrix factorization* of both P and S as

$$X \approx PS. \tag{3.12}$$

One aims in finding *meaningful* matrices P and S such that (3.12) holds.

The connection to data compression arises if one assumes that it is given a (hyperspectral) datacube X and one seeks a compressed representation of it in the form of a matrix factorization (3.12). If it is demanded that the dimension ρ is very small compared to the other two, i.e. $\rho \ll \min\{n, m\}$, then the matrices P and S are both together significantly smaller than X. As it will be shown in Chapter 6 on the example of hyperspectral mass spectrometry data, this form of compression jointly covers both of the redundancy aspects mentioned in Section 3.1, namely the spatial and the spectral one.The idea is that the columns of P and the rows of S will in some sense form a basis of the whole data X.

Remark 3.3.6. In the special case that all elements of P and S (and X) are nonnegative, (3.12) is referred to as *nonnegative matrix factorization*. This will be studied in Section 6.4 with the application to mass spectrometry data.

Remark 3.3.7. The form in (3.12) is basically of the same type as in (3.4). This can be motivated as follows: Until now, it was assumed that Ψ in (3.4) is a known dictionary. Now, assume that also $\Psi \in \mathbb{R}^{n \times \rho}$ is an *unknown* linear operator. Then, the task is to reconstruct both unknowns the (convolution) operator Ψ and the coefficients λ under the knowledge of the (convoluted) signal y, so that (3.4) is fulfilled. This problem is known as the *blind deconvolution* [180]. If $P = \Psi$ and $m = 1$, (3.12) becomes the just mentioned (one dimensional) blind deconvolution problem (3.4).

As before, a potential way to find matrices P and S such that (3.12) holds is by considering the following minimization problem

$$\underset{P \in \mathbb{R}^{n \times \rho}, S \in \mathbb{R}^{\rho \times m}}{\operatorname{argmin}} \frac{1}{2} \|PS - X\|_F^2. \tag{3.13}$$

The term $\frac{1}{2}\|PS - X\|_2^2$ is still convex in either P or S, but not necessarily jointly convex which implies the possibility of local minima, see Example 3.3.8. Therefore solving the factorization problem (3.12) is difficult in general and it is necessary to use as much a priori information as possible.

Example 3.3.8 (Counterexample for the joint convexity). Fix $X = 1$ and define $f : \mathbb{R}^2 \to \mathbb{R}_+$ as $f(P,S) := \frac{1}{2}\|PS - 1\|_F^2 = \frac{1}{2}(PS - 1)^2$. Following the definition of a convex function given in (2.7) on page 18 fix $\lambda = \frac{1}{2}$ and the two coordinates $(P,S) = (1,3)$ as well as $(\tilde{P}, \tilde{S}) = (3,1)$. Then it holds

$$f\left(\frac{1}{2}(1,3) + \frac{1}{2}(3,1)\right) = f(2,2) = \frac{9}{2} \nleq \frac{1}{2}f(1,3) + \frac{1}{2}f(3,1) = 2.$$

Therefore, the function f is not (jointly) convex.

The often used strategy to solve (3.13) is to do an alternate minimization, meaning that one iteratively updates P for fixed S and then S for fixed P, as it has been popularized by Lee and Seung [120, 121], Hoyer and Dayan [108] and Donoho and Stodden [75].

Lee and Seung, for instance, have shown in [120] that the Euclidean distance (3.13) is non increasing under the multiplicative update rule

$$P_{k+1} = P_k \circ \frac{X S_{k+1}^T}{P_k S_{k+1} S_{k+1}^T}, \quad S_{k+1} = S_k \circ \frac{P_{k+1}^T X}{P_{k+1}^T P_{k+1} S_k}. \tag{3.14}$$

Here, both the $\circ$ and the fraction denote the component-wise multiplication and division of the expressions.

Without any further information or restrictions to the functional the solution, i.e. the multiplication PS such that (3.12) holds, is clearly not unique. In fact, for any invertible matrix $Q \in \mathbb{R}^{\rho \times \rho}$ also $\bar{P} = PQ^{-1}$ and $\bar{S} = QS$ fulfill $X \approx \bar{P}\bar{S} = PQ^{-1}QS$. For this reason, geometric conditions to guarantee uniqueness are studied in [75]. Another approach that already appeared several times in this work is to add constraints $\Pi_1(P)$ and $\Pi_2(S)$ to the functional (3.13) such as sparsity constraints [75, 108],

$$\underset{P \in \mathbb{R}^{n \times \rho}, S \in \mathbb{R}^{\rho \times m}}{\operatorname{argmin}} \frac{1}{2} \|PS - X\|_F^2 + \alpha \Pi_1(P) + \beta \Pi_2(S). \tag{3.15}$$

For the minimization of this functional there exist several algorithms. One is the PPXA (cf. Section 2.2) as one from the proximal splitting methods [60, 127]. It first splits the given problem in several smaller problems and then solves them separately. Another way is the *alternating direction method of multipliers* (ADMM) [82, 196]. There, the optimization problem is first formulated as a Lagrangian function and then minimized with respect to all its arguments. Very similar to ADMM are *surrogate* or *auxiliary function* approaches where the functional to be minimized is replaced by a closely related but still different (surrogate) functional for analytical and numerical advantages, see e.g. [66] and [135] and the references therein.

Another idea studied in [157] is to solve (3.15) via a fixed point iteration in its unconstrained form while ensuring the non-negativity. This is accomplished by adding a positivity preserving function with a specific sequence $\{c_k\}$ to the functional (3.15). For details regarding these methods, refer to the resources cited above.

Within the whole chapter the full data was available to compress. As already motivated in Chapter 1 it does not seem to be reasonable to measure all the data first and then neglect the redundancies afterwards. Instead, applying the compression in the acquisition process seems more efficient, leading to the field of compressed sensing which will be introduced in detail in the next chapter.

4 | Compressed Sensing

Within this chapter a recently developed acquisition paradigm called *compressed sensing* will be studied. As the name already indicates, instead of measuring the full data first and then removing most of its redundant information, it applies a compression step directly during the sensing procedure. The presentation of the results in this chapter will follow those from the textbooks [86, 93, 171, 178] as well as from the article [37].

4.1. Introduction

The combination of classical Shannon-Nyquist [152, 175] sampling and compression steps is one of the main ideas of compressed sensing. It turns out that it is possible to represent or reconstruct data using sampling rates much lower than the Nyquist rate [47, 49, 72].

One of the key concepts is the notion of sparsity. A signal $x \in \mathbb{R}^n$ is called *sparse*, if most of its entries are equal to zero, that is, if its support $\|x\|_0$ is of cardinality $k \ll n$. An *s-sparse* signal is one for which maximal s samples have a nonzero value, i.e. $\|x\|_0 \leqslant s$. The set of all s-sparse signals is described by

$$\Sigma_s := \{x \in \mathbb{R}^n : \|x\|_0 \leqslant s\},$$

with which the *best s-term approximation error* of a vector $x \in \mathbb{R}^n$ in ℓ_p can be defined as

$$\sigma_s(x)_p := \inf_{z \in \Sigma_s} \|x - z\|_p. \tag{4.1}$$

Usually, real data is in fact not exactly sparse. It is rather *compressible* in the sense that $\sigma_s(x)_p$ decays quickly in s. The value $\sigma_s(x)_p$ constitutes the best, i.e. the smallest error possible if an s-sparse approximation z of the given vector x is considered. The infimum of (4.1) is given by any vector z whose s entries equal the s largest entries in magnitude of x. Therefore, the error $\sigma_s(x)_p$ is described by the left $n - s$ coefficients of x. For this, one defines the *nonincreasing rearrangement* of x by

$$x_1^* \geqslant x_2^* \geqslant \ldots \geqslant x_n^* \geqslant 0,$$

where $x_j^* = |x_{\pi(j)}|$, $j \in [n]$ and $\pi : [n] \to [n]$ is a corresponding permutation on the index set $[n] := \{1, \ldots, n\}$, from which one follows that

$$\sigma_s(x)_p = \left(\sum_{j=s+1}^{n} (x_j^*)^p \right)^{1/p}, \qquad 0 < p < \infty. \tag{4.2}$$

As the next Lemma states, it is very useful to study ℓ_q balls with small q (e.g. $q \leqslant 1$) in the context of sparse or compressible signals, as this tends to a potentially small best s-term approximation error.

Lemma 4.1.1 (Section 6.3.2, [171]). *Let $0 < q < p \leqslant \infty$ and set $r = 1/q - 1/p > 0$. Then for all $s = 1, \ldots, n$ and $x \in \{z \in \mathbb{R}^n : \|z\|_q \leqslant 1\}$, the following power law holds:*

$$\sigma_s(x)_p \leqslant s^{-r}.$$

Proof. Let S be the set of indices of the s-largest magnitude entries of x and $x^* \in \mathbb{R}^n$ be the nonincreasing rearrangement of x. By taking into account the identity (4.2) as well as the fact that $s(x_s^*)^q \leqslant \sum_{j=1}^{s}(x_j^*)^q$, it follows

$$\sigma_s(x)_p^p = \sum_{j=s+1}^{n} (x_j^*)^p \leqslant (x_s^*)^{p-q} \sum_{j=s+1}^{n} (x_j^*)^q$$

$$\leqslant \left(\frac{1}{s} \sum_{j=1}^{s} (x_j^*)^q \right)^{\frac{p-q}{q}} \left(\sum_{j=s+1}^{n} (x_j^*)^q \right)$$

$$\leqslant \left(\frac{1}{s} \|x\|_q^q \right)^{\frac{p-q}{q}} \|x\|_q^q \leqslant s^{-\frac{p-q}{q}}.$$

Taking the power $1/p$ on both sides completes the proof. $\qquad\square$

Motivated by this result the general compressed sensing setting will now be studied. There, one aims to perform compression during the acquisition process, rather than measuring the full information of a signal and compressing it afterwards. More formally, given a signal or data $x \in \mathbb{R}^n$, it suffices to take only $k = 1, \ldots, m \ll n$ linear measurements $y_k \in \mathbb{R}$ using linear test functions $\hat{\varphi}_k \in \mathbb{R}^n$, i.e.

$$y_k = \langle \hat{\varphi}_k, x \rangle.$$

In matrix notation this reads

$$y = \Phi x, \tag{4.3}$$

where $\Phi \in \mathbb{R}^{m \times n}$ is called the *measurement matrix* and has rows filled with the functions $\hat{\varphi}_k$. As $m \ll n$, this system is highly underdetermined. In terms of inverse problems [87, 129, 161] this problem is an *ill-posed* inverse problem, as the linear operator Φ is not injective, leading to infinitely many solutions x of (4.3). Without any additional knowledge on the signal x, its reconstruction from the measurements y is hopeless. As it will be shown in the following sections, adding the information about the sparsity of a given signal x changes the situation positively.

4.2. Uniqueness, sparseness and other properties

As described in Section 3.3, if one wants to solve (4.3) and if one knows that the acquired signal x is sparse, a natural way to go is to solve the minimization problem

$$\underset{z \in \mathbb{R}^n}{\operatorname{argmin}} \|z\|_0 \text{ subject to } \Phi z = y, \tag{4.4}$$

Besides all the possible s-sparse solutions of (4.4), it would be desirable to find conditions that guarantees that there is in fact only one. For the rest of this section the following is a simple but important observation.

Lemma 4.2.1 ([93, Section 2.2]). *Let $x \in \mathbb{R}^n$ be s-sparse. Then the following statements are equivalent:*

a) x is the unique s-sparse solution to $y = \Phi z$ with $y = \Phi x$, that is,

$$\{z \in \mathbb{R}^n \mid Az = Ax, \|z\|_0 \leqslant s\} = \{x\}.$$

b) x can be reconstructed as the unique solution of (4.4).

Proof. The implication b) $\Rightarrow$ a) is clear. For the implication a) $\Rightarrow$ b) let x be the unique s-sparse solution to $y = \Phi z$ with $y = \Phi x$. One of the solutions x° of (4.4) is s-sparse and satisfies $Ax^\circ = y$, so that $x^\circ = x$. □

Remark 4.2.2. In the following, conditions are described guaranteeing a unique sparse solution of (4.3), i.e. of $y = \Phi x$ where $\Phi \in \mathbb{R}^{m \times n}$ is a measurement matrix with $m \ll n$. The presented results should always be read with having Lemma 4.2.1 in mind. Wherever it is described under which conditions there exists a unique solution, it also means that under these unique recovery using the ℓ_0 minimization problem (4.4) is possible.

As it will be described in the following, one of those conditions is given via the *null space* of Φ, denoted by

$$\mathcal{N}(\Phi) := \{x \in \mathbb{R}^n : \Phi x = 0\}.$$

Let x° satisfy $\Phi x^\circ = y$ and let $\eta \in \mathcal{N}(\Phi)$ be any element in the null space of Φ. It is then clear that any other solution to $\Phi x = y$ has the form $x = x^\circ + \eta$.

Now, let us suppose that there are two distinct s-sparse solutions x° and $x^\sharp$, i.e. $x^\circ, x^\sharp \in \Sigma_s$. At first it is clear that $\Phi(x^\circ - x^\sharp) = 0$, because of which $x^\circ - x^\sharp \in \mathcal{N}(\Phi)$. As x° and $x^\sharp$ are different, their subtraction is not the zero vector but one that lies in the space Σ_{2s}, i.e. $x^\circ - x^\sharp \in \Sigma_{2s}$. In this way one has found a first condition on the uniqueness of a s-sparse solution: If $\Phi x = y$ has more than one s-sparse solution, the null space $\mathcal{N}(\Phi)$ must contain a nonzero $2s$-sparse vector, or said in a contrapositive way:

Lemma 4.2.3 (Uniqueness via the null space [37, Lemma 2.1]). *Suppose that $\Sigma_{2s} \cap \mathcal{N}(\Phi) = \{0\}$, that is, all nonzero elements in the null space of Φ have at least $2s + 1$ nonzero components. Then any s-sparse solution to $\Phi x = y$ is unique.*

The next Lemma is a simple reformulation of Lemma 4.2.3 and states a possible way to check whether Φ fulfills $\Sigma_{2s} \cap \mathcal{N}(\Phi) = \{0\}$ or not. It also forms the basis for the subsequent considerations.

Lemma 4.2.4 ([37, Lemma 2.2]). *The condition* $\Sigma_{2s} \cap \mathcal{N}(\Phi) = \{0\}$, *holds if and only if every subset of* $2s$ *columns of* Φ *is linearly independent.*

A term that often arises in this context is the *spark* of a given matrix, which was introduced in [73]:

Definition 4.2.5. The *spark* of a given matrix Φ is the smallest number of columns of Φ that are linearly dependent.

With this notion, Lemma 4.2.3 and Lemma 4.2.4 can be combined in the following theorem:

Theorem 4.2.6 (Uniqueness via the spark [73, Corollary 1]). *For any vector* $y \in \mathbb{R}^m$, *there exists at most one signal* $x \in \Sigma_s$ *such that* $y = \Phi x$ *if and only if* spark$(\Phi) > 2s$.

As it is clear that spark$(\Phi) \in [2, m+1]$ if there are no zero columns in Φ, this theorem yields a requirement on the measurements, namely $m \geqslant 2s$.

Consider now the two situations that one either has to build a measurement matrix Φ or that a matrix Φ is given and one has to check if the condition $\Sigma_{2s} \cap \mathcal{N}(\Phi) = \{0\}$ is fulfilled. In fact, this is impracticable, especially for large n, m and s, as it is required to check all $\binom{n}{2s}$ subsets of columns of Φ. In fact it is shown in [189] that determining the spark of a given matrix is NP-hard. Therefore, as this brute force variant is not feasible at all, an alternative is required that is easier to check.

4.2.1. Coherence

A simpler way to measure the suitability of the measurement matrix Φ is the so-called *coherence* between its columns:

Definition 4.2.7 (Coherence). Let $\Phi \in \mathbb{R}^{m \times n}$ be a matrix with columns $\varphi_1, \ldots, \varphi_n$. The *coherence* $\mu(\Phi)$ of the matrix Φ is defined as

$$\mu(\Phi) := \max_{1 \leqslant i < j \leqslant n} \frac{|\langle \varphi_i, \varphi_j \rangle|}{\|\varphi_i\|_2 \|\varphi_j\|_2}. \tag{4.5}$$

Within compressed sensing one is interested in small coherence or *incoherence*, respectively. This indicates that the largest correlation or dependency between the columns of Φ, i.e. the largest value in magnitude of the inner product in (4.5) is small. To see this first note that it holds $\mu(\Phi) \leqslant 1$ due to the Cauchy-Schwarz inequality. For the lower bound, which is also known as the Welsh bound [206], one can show that $\mu(\Phi) \geqslant \sqrt{\frac{n-m}{m(n-1)}}$, see e.g. [93, Theorem 5.7] for a proof. Therefore, it holds

$$\mu(\Phi) \in \left[\sqrt{\frac{n-m}{m(n-1)}}, 1\right]. \tag{4.6}$$

However, if the number of taken measurements is much smaller than the signal dimension n, i.e. $m \ll n$, then the lower bound is approximately $1/\sqrt{m}$. Now, in light of Theorem 4.2.6, the following lemma can be shown:

Lemma 4.2.8 ([86, Lemma 1.4]). *For any matrix Φ it holds*

$$\mathrm{spark}(\Phi) \geqslant 1 + \frac{1}{\mu(\Phi)}.$$

Proof. cf. [86] or [73]. □

In combination with Theorem 4.2.6, another condition on the measurement matrix Φ is given that guarantees uniqueness of the s-sparse solution of $\Phi x = y$:

Theorem 4.2.9 ([73, Theorem 7]). *If it holds*

$$s < \frac{1}{2}\left(1 + \frac{1}{\mu(\Phi)}\right), \tag{4.7}$$

then for each measurement vector $y \in \mathbb{R}^m$ there exists at most one signal $x \in \Sigma_s$ such that $y = \Phi x$.

As said, measurement matrices with small coherences are of interest in compressed sensing. To clarify this, let us study the results in Theorem 4.2.9 with an example.

Example 4.2.10 (Sparsity vs. measurements with Theorem 4.2.9). Let $x \in \mathbb{R}^n$ with $n = 100$ be a given signal and apply $m = 20$ measurements via $\Phi x = y \in \mathbb{R}^m$. Due to (4.6), in the worst case the coherence $\mu(\Phi)$ is 1, leading to $s < 1$ in (4.7). As $s \in \mathbb{N}$, this is surely never true, meaning that there is probably more than one sparse solution $x \in \Sigma_s$ to $\Phi x = y$. On the other hand, the lower bound of the coherence becomes $1/\sqrt{m}$ as $n \gg m$, leading to $s = \mathcal{O}(\sqrt{m})$[1]. Therefore, the inequality (4.7) reads

$$ s < \frac{1}{2}\left(1 + \sqrt{m}\right) = \frac{1}{2}\left(1 + \sqrt{20}\right) \approx 2.7. $$

In words, this means that in this case of lowest coherence the signal sparsity should not exceed $s = 2$ to ensure uniqueness of the desired solution of $\Phi x = y$. Let us consider the problem from the other side, namely that one knows a priori that the sparsity level is $s = 5$. In this way, one can get an idea of how many measurements m are theoretically needed to ensure a unique solution. Since there is no information on the relation between n and m available it is reasonable to use the exact lower bound in (4.6). Note that here it is again assumed to have lowest coherence of Φ possible. Starting from (4.7) then leads to the relation

$$ s = 5 < \frac{1}{2}\left(1 + \frac{1}{\mu(\Phi)}\right) \quad \Longleftrightarrow \quad \sqrt{\frac{100 - m}{99m}} < \frac{1}{9}, $$

following the condition $m > 45$.

It can be seen that the less sparse the measured signal is, the more measurements are needed for a unique and possibly good reconstruction. This example more generally illustrates what was already recognizable in the coin example in Section 1.3: If one wants to solve the linear inverse problem $\Phi x = y$, the underlying parameter dimensions, the sparsity level s of the signal x as well as the number of taken measurements m in Φ go hand in hand.

4.2.2. Restricted isometry property

Despite the fact that the coherence is easy to calculate, it is not nearly as often used as a reconstruction criterion as the so-called *restricted isometry*

---

[1] $f(h) = \mathcal{O}(g(h)) :\Longleftrightarrow \exists C > 0 : |f(h)| \leqslant C|g(h)|.$

property (RIP). A matrix $\Phi \in \mathbb{R}^{m \times n}$ fulfills the RIP of order s if there exists a constant $\delta \in (0,1)$ such that for all $x \in \Sigma_s$ it holds

$$(1 - \delta)\|x\|_2^2 \leqslant \|\Phi x\|_2^2 \leqslant (1 + \delta)\|x\|_2^2.$$

The RIP enjoys its popularity due to the fact that one can easily show that it is satisfied for a wide variety of matrices with some overwhelming probability [47, 49, 93]. Therefore, in the following the restricted isometry property is motivated and introduced.

In the following, let $\bar{x} = x/\|x\|_2$. This is less of a limitation to ℓ_2 unit vectors, but rather a simplification in the presentation. Then it is clear that if $x \neq 0$, then $\Phi x = 0$ if and only if $\Phi \bar{x} = 0$.

As a reminder, for a unique s-sparse solution to $\Phi x = y$ both Lemma 4.2.3 and Lemma 4.2.4 required that no nonzero vector $x \in \Sigma_{2s}$ lies in the null space of Φ, i.e. $\Sigma_{2s} \cap \mathcal{N}(\Phi) = \{0\}$. In terms of ℓ_2 unit norm vectors $\bar{x}$ this means that no $\bar{x} \in \Sigma_{2s}$ satisfies $\Phi \bar{x} = 0$. One canonical approach is to require a positive constant $c_1 > 0$ such that $\|\Phi \bar{x}\|_2^2 \geqslant c_1$ for all $u \in \Sigma_{2s}$. Thus, if one can ensure the existence of such a constant c_1, the condition $\Sigma_{2s} \cap \mathcal{N}(\Phi) = \{0\}$ is fulfilled.

Next note that the mapping $x \mapsto \|\Phi x\|_2^2$ from $\mathbb{R}^n$ to $\mathbb{R}$ is continuous and the set of all $2s$-sparse vectors in $\mathbb{R}^n$ is compact [37]. Hence, due to the extreme value theorem there must exists a constant $c_2 > 0$ such that $\|\Phi \bar{x}\|_2^2 \leqslant c_2$. Therefore, for all $\bar{x} \in \Sigma_{2s}$ it holds the following chain of inequalities:

$$c_1 \leqslant \|\Phi \bar{x}\|_2^2 \leqslant c_2. \tag{4.8}$$

For the next let c be a nonzero scalar and let Φ fulfill $\Sigma_{2s} \cap \mathcal{N}(\Phi) = \{0\}$. Then one easily checks that it holds $\Sigma_{2s} \cap \mathcal{N}(\Phi') = \{0\}$ for $\Phi' = c\Phi$. Therefore one is free to multiply (4.8) by $2/(c_1 + c_2)$. With the definitions $\delta := (c_2 - c_1)/(c_1 + c_2)$ as well as $\Phi := \sqrt{2/(c_1 + c_2)}\Phi$, (4.8) becomes

$$1 - \delta \leqslant \|\Phi \bar{x}\|_2^2 \leqslant 1 + \delta.$$

Under the assumption that Φ acts in general only nearly as an isometry it is $0 < c_1 < c_2$ and thus $\delta \in (0,1)$. This finally leads to the following definition of the restricted isometry property.

Definition 4.2.11 (Restricted isometry property). The matrix $\Phi \in \mathbb{R}^{m \times n}$ satisfies the *restricted isometry property* (RIP) of order s if there exists some constant $\delta \in (0, 1)$, such that

$$(1 - \delta)\|x\|_2^2 \leqslant \|\Phi x\|_2^2 \leqslant (1 + \delta)\|x\|_2^2 \tag{4.9}$$

for all s-sparse vectors $x \in \Sigma_s$. The smallest constant δ for which this inequality chain holds is called the *restricted isometry (property) constant* and denoted by $\delta = \delta_s$.

Since $\Sigma_s \subset \Sigma_{s+1}$, it is clear that the sequence of the restricted isometry constants is nondecreasing:

$$\delta_1 \leqslant \delta_2 \leqslant \ldots \leqslant \delta_s \leqslant \delta_{s+1} \leqslant \ldots \leqslant \delta_n.$$

It was shown in [189] that finding the smallest constant δ_s for a given matrix Φ is NP-hard. One can further show that with Definition 4.2.11, Lemma 4.2.3 can be rewritten with the notion of the RIP in the following way:

Lemma 4.2.12 (Uniqueness via the RIP [37, Lemma 2.5]). *If a matrix Φ satisfies the RIP of order $2s$ for some $s \geqslant 1$, then $\Sigma_{2s} \cap \mathcal{N}(\Phi) = \{0\}$ and any s-sparse solution to $\Phi x = y$ is unique.*

Remark 4.2.13. The condition $\Sigma_{2s} \cap \mathcal{N}(\Phi) = \{0\}$ can also be understood in such a way that the linear operator Φ is injective on the set of all $2s$-sparse vectors, i.e. on Σ_{2s}. A closely related concept is given by the so-called *finite basis injectivity* (FBI) [35, 68] which is defined as follows:

Let $I \subset \mathbb{N}$, define the sequence space $\ell^2 = \{(x_i)_{i \in I} \subset \mathbb{R} \mid \sum_{i \in I} |x_i|^p < \infty\}$ and let $\mathcal{H}$ be a Hilbert space. An operator $K : \ell^2 \to \mathcal{H}$ has the FBI property, if for all finite subsets $S \subset \mathbb{N}$ the operator $K|_S$ is injective, i.e. $Ku = Kv$ and $u_k = v_k = 0$ for all $k \notin S$ implies $u = v$.

Operators which fulfill the RIP of order s also obey the FBI property with respect to all subsets $T \subset S$ with $|S| \leqslant s$. Contrariwise, not every FBI operator also satisfies the RIP. For this to see it is enough to study the counterexample $K : \mathbb{R}^2 \to \mathbb{R}^2$ with

$$K = \begin{pmatrix} 2 & 0 \\ 0 & 3 \end{pmatrix}.$$

K is clearly injective and is invertible with inverse K^{-1}. K has the smallest eigenvalue $\lambda_{\min} = 2$ and the largest eigenvalue $\lambda_{\max} = 3$. It holds $1/\|K^{-1}\|_2\|x\|_2 \leqslant \|Kx\|_2 \leqslant \|K\|_2\|x\|_2$ with operator norms $\|K^{-1}\|_2 = 1/\lambda_{\min} = 1/2$ and $\|K\|_2 = \lambda_{\max} = 3$. In light of the RIP (4.9), there is no $\delta \in (0,1)$ such that the RIP is fulfilled. In other words: K does not fulfill the RIP of order $s = 2$. Hence, the RIP is a more strict requirement on the operator.

Remark 4.2.14. At the time of this thesis, analysis of CS in infinite dimensions is not as extensive done as it is for the finite dimensional case. The just discussed FBI, however, is one of the known tools to formulate exact recovery conditions in the first case and is used in e.g. [35, 68, 105, 191]. For completeness it should be mentioned that A. C. Hansen *et al.* have motivated that the typical definitions of sparsity and coherence should probably be replaced by *asymptotic* variants of them, see [4, 31].

The name RIP comes from the fact that the restrictions of Φ to certain subspaces are isometries. If there are two s-sparse elements $x, x' \in \Sigma_s$, then $x - x' \in \Sigma_{2s}$ and it follows

$$\|\Phi x - \Phi x'\|_2 = \|\Phi(x - x')\|_2 \geqslant \sqrt{1 - \delta_{2s}}\|x - x'\|_2. \qquad (4.10)$$

If δ_{2s} goes to zero, Φ acts more and more like an isometry on Σ_s, which implies that the vectors Φx and $\Phi x'$ are more easily distinguishable. Hence, from that point of view one is interested in the smallest possible isometry constant. The restriction to have $\delta_{2s} < 1$ (or generally $\delta < 1$) leads to $\|\Phi x - \Phi x'\|_2 > 0$ for all distinct s-sparse vectors $x, x' \in \mathbb{R}^n$ and is to ensure that those yield different measurement vectors.

The following small example is intended to show how a RIP constant for a matrix can be determined for a certain sparsity level.

Example 4.2.15 (RIP constant for a specific matrix [37, Exercise 11]). Let $\Phi = [1/2 \; 4/3] \in \mathbb{R}^{1 \times 2}$. First, Φ fulfills the RIP of order $s = 1$ with constant $\delta_1 = 7/9 < 1$. For this to show let $x = (x_1 \; x_2)^T \in \Sigma_1 \subseteq \mathbb{R}^2$ be any 1-sparse vector. Then it is

$$\|\Phi x\|_2^2 \leqslant (4/3)^2\|x_1 + x_2\|_2^2 = 16/9(\|x_1\|_2^2 + \|x_2\|_2^2) = (1 + 7/9)\|x\|_2^2.$$

Contrariwise, with the same 1-sparse argument it holds

$$\|\Phi x\|_2^2 \geqslant 1/4\|x\|_2^2 > 2/9\|x\|_2^2 = (1 - 7/9)\|x\|_2^2.$$

Second, Φ does not fulfill the RIP of order 2, as the following example shows: If $x = (2 \; 6)^T$, then $\|\Phi x\|_2^2 = 81$, but $\|x\|_2^2 = 40$. Since it is required to have a constant $\delta_2 \in (0, 1)$, there does not exist such δ_2 so that $\|\Phi x\|_2^2 = 81 \leqslant (1 + \delta_2)40 = (1 + \delta_2)\|x\|_2^2$ holds, as $\delta_2 > 1$ is required.

The RIP can be easily connected with the notion of the spark of a given matrix. As described above, one of the inequalities in the definition of the RIP follows from the motivation to have no nonzero vector $x \in \Sigma_{2s}$ in the null space of Φ. If Φ now has the RIP of order $2s$ with constant $\delta_{2s} > 0$, then every set of $2s$ columns of Φ is linearly independent, which implies the condition spark$(\Phi) > 2s$ of Theorem 4.2.6. In addition, the RIP can be connected to coherence from Definition 4.2.7 via the following lemma.

Proposition 4.2.16 (RIP and coherence [40, Proposition 4.1]). *If Φ has unit-norm columns and coherence $\mu(\Phi)$, then Φ satisfies the RIP of order s with constant*

$$\delta_s \leqslant (s - 1)\mu(\Phi).$$

Proof. Let $x \in \mathbb{R}^n$ be s-sparse and assume w.l.o.g. that supp$(x) = \{1, \ldots, s\}$. Then it holds

$$\|\Phi x\|_2^2 = \sum_{i,j}^{s} \langle \varphi_i, \varphi_j \rangle x_i x_j = \|x\|_2^2 + \sum_{1 \leqslant i \neq j \leqslant s} \langle \varphi_i, \varphi_j \rangle x_i x_j.$$

Applying the definition of the coherence, the second term can be bounded by

$$\left| \sum_{1 \leqslant i \neq j \leqslant s} \langle \varphi_i, \varphi_j \rangle x_i x_j \right| \leqslant \mu(\Phi) \sum_{1 \leqslant i \neq j \leqslant s} |x_i x_j|$$

$$\leqslant \mu(\Phi)(s - 1) \sum_{i=1}^{s} |x_i|^2$$

$$= \mu(\Phi)(s - 1)\|x\|^2.$$

This leads to the chain of inequalities

$$\left(1 - (s-1)\mu(\Phi)\right)\|x\|_2^2 \leqslant \|\Phi x\|_2^2 \leqslant \left(1 + (s-1)\mu(\Phi)\right)\|x\|_2^2,$$

and therefore

$$\delta_s \leqslant (s-1)\mu(\Phi),$$

i.e. a bound for the RIP constant δ_s for the matrix Φ. $\qquad\square$

As previously mentioned, the RIP enjoys its high popularity to the fact that many matrices satisfy it with high probability. Apart from random matrices, which will be discussed next, there also exist other deterministic ones that fulfill either the condition $\Sigma_{2s} \cap \mathcal{N}(\Phi) = \{0\}$ or the RIP, cf. Lemma 4.2.3 and Lemma 4.2.12.

One of them is the $m \times n$ Vandermonde matrix V, constructed from n distinct scalars $t_1, t_2 \ldots, t_n$,

$$V = \begin{pmatrix} 1 & 1 & 1 & \cdots & 1 \\ t_1 & t_2 & t_3 & \cdots & t_n \\ t_1^2 & t_2^2 & t_3^2 & \cdots & t_n^2 \\ \vdots & \vdots & \vdots & \ddots & \vdots \\ t_1^{m-1} & t_2^{m-1} & t_3^{m-1} & \cdots & t_n^{m-1} \end{pmatrix}.$$

It can be shown that all its $2s \times 2s$ minors (itself Vandermonde matrices) are invertible provided that all n scalars are distinct and strict positive [89]. The largest such minor matrix is given for $m = 2s$. Hence, V has $\mathrm{spark}(V) = m + 1$ [58], but unfortunately V is poorly conditioned for large n which causes numerical instabilities solving $Vx = y$ [96].

In [69] it is presented a way to construct deterministic compressed sensing matrices of size $m \times n$ that fulfill the RIP of order

$$s = \mathcal{O}(\sqrt{m}\log(m)/\log(n/m)).$$

The problem here is that these matrices require

$$m = \mathcal{O}(s^2 \log(n)) \tag{4.11}$$

measurements which make this approach unusable for practical values of s and n. In particular, the factor s^2 has a big impact when the underlying signal is not very sparse. For example, if the number of non-zero entries are doubled, then one needs to quadruple the number of measurements.

For more examples on non-random matrices that are, in principle, able to be used in compressed sensing, refer to [80] and [93] and the references therein.

The result that random matrices fulfill the RIP with high probability is given in the next Theorem 4.2.17.

Theorem 4.2.17 (RIP for random matrices [17, 49, 171]). *Let $x \in \Sigma_s$ and assume $\Phi \in \mathbb{R}^{m \times n}$ to be a random matrix satisfying the concentration property[2]*

$$\mathbb{P}\big(\big|\|\Phi x\|_2^2 - \|x\|_2^2\big| \geq t\|x\|_2^2\big) \leq 2\exp(-c_0 t^2 m), \qquad (4.12)$$

for $0 < t < 1$ and some constant $c_0 > 0$. Then there exists a constant C depending only on c_0 such that the restricted isometry constant δ_s of Φ satisfies $\delta_s \leq \zeta \in (0, 1)$ with probability exceeding $1 - \varepsilon$ provided

$$m \geq C\zeta^{-2}\left(s\ln\left(\frac{n}{m}\right) + \ln(\varepsilon^{-1})\right). \qquad (4.13)$$

Proof. cf. [17] or [37]. $\qquad\qquad\qquad\qquad\qquad\qquad\qquad\qquad\square$

Remark 4.2.18. The expression $\big|\|\Phi x\|_2^2 - \|x\|_2^2\big| \geq t\|x\|_2^2$ given in (4.12) is just another form of the RIP. Hence, the concentration inequality bounds the probability that a specific matrix Φ satisfies the RIP.

Typical CS results require that a matrix Φ from a certain distribution fulfills the RIP with some RIP constant δ_s. Theorem 4.2.17 states how many measurements m are needed given system parameters so that the matrix will indeed fulfill the RIP with high probability. If the measurements are done at random, (4.13) implies that the larger ζ (and with that δ_s), the smaller is the number of measurements required to obtain a high probability $1 - \varepsilon$ to fulfill the RIP.

Random matrices $\Phi \in \mathbb{R}^{m \times n}$ that satisfy the concentration inequality (4.12) are in particular Gaussian and Bernoulli matrices [2, 3]. The

---

[2]$\mathbb{P}(\cdot)$ denotes a probability measure on (Ω, Σ), where Σ denotes a σ-algebra on the sample space Ω. The probability of an event $B \in \Sigma$ is given by $\mathbb{P}(B) = \int_B d\mathbb{P}(\omega) = \int_\Omega I_B(\omega)d\mathbb{P}(\omega)$ where $I_B(\cdot)$ is the characteristic function.

entries in a Gaussian matrix are independent standard Gaussian random variables with expectation 0 and variance $1/m$. In addition, the entries in a Bernoulli matrix are independent Rademacher variables, meaning that they take their values ± 1 with equal probability. Together with Lemma 4.2.12, the resulting inequality (4.13) from Theorem 4.2.17 can be interpreted in such a way that unique recovery of the signal via ℓ_1 minimization is given with high probability, provided

$$m \geqslant Cs \log\left(\frac{n}{m}\right) \geqslant Cs \log\left(\frac{n}{s}\right), \tag{4.14}$$

as $s \leqslant m$ from Theorem 4.3.1. An often mentioned heuristic magnitude for the number of needed measurements m for ℓ_1 minimization in (4.19) to recover a s-sparse signal $x \in \mathbb{R}^n$ is given by $m \geqslant 4s$, see for instance [48] and [50].

Theorem 4.2.17 states that any random matrix that fulfills the concentration inequality (4.12) has the RIP with constant $\delta_s \leqslant \zeta$. To understand the connection on the wanted small isometry constants on the one side and the increasing of their value on the other side note the following Theorem 4.2.19. It details the results from Theorem 4.2.17 for Gaussian matrices and yields a bound for the restricted isometry constant δ_s.

Theorem 4.2.19 (RIP for Gaussian matrices [93]). *Let* $\Phi \in \mathbb{R}^{m \times n}$ *with* $m < n$ *be a Gaussian matrix. For* $\eta, \varepsilon \in (0,1)$, *assume that*

$$m \geqslant 2\eta^{-2}\left(s \ln\left(e\frac{n}{s}\right) + \ln\left(2\varepsilon^{-1}\right)\right). \tag{4.15}$$

Then with probability of at least $1 - \varepsilon$ *the restricted isometry constant* δ_s *of* $\frac{1}{\sqrt{m}}A$ *satisfies*

$$\delta_s \leqslant 2\left(1 + \frac{1}{\sqrt{2\ln(e\frac{n}{s})}}\right)\eta + \left(1 + \frac{1}{\sqrt{2\ln(e\frac{n}{s})}}\right)^2 \eta^2. \tag{4.16}$$

Proof. cf. [93]. □

In light of Lemma 4.2.12 and the comment thereafter, it is known that $\delta_{2s} < 1$ is sufficient for a matrix to ensure unique recovery of any s-sparse signal in the noiseless case. If one wants to have also robust reconstruction of the solution in the presence of noise, it will be seen in Theorem 4.3.2 that it is required to have isometry constants less than a certain value, here $\delta_{2s} < \sqrt{2} - 1 \approx 0.4142$. In literature one often reads that larger restricted isometry constants such as $\delta_{2s} < 2/(3 + \sqrt{2}) \approx 0.4531$ from [92] are *improvements*, which seems to be a contradiction to the above mentioned goal to have smallest possible constants to be close to an isometry. The reason for increasing the constant is that one aims at guaranteeing that more (random) matrices fulfill the RIP so that those measurement matrices are sufficient in practice for stable and exact recovery. A matrix that fulfills the RIP with $\delta_s = 0.3$ will also fulfill the RIP with $\delta_s = 0.5$, but not vice versa.

If the right hand side in (4.16) is large, it implies a large η, and therefore a smaller value m in (4.15) for the number of measurements required. However, this is of course at the expense of the isometry property itself.

Remark 4.2.20. It is at this point worth mentioning a nice visualization of the relations between the parameters n, m and s. As Donoho and Tanner presented in [77] among others in the context of CS, a possible approach to visualize those relations is by plotting the fractions $\tau = m/n$ and the level of sparsity $\rho = s/m$ against each other. For illustration take the coin problem from the Section 1.3 for $n = 100$ fixed coins and change the values for the sparsity level s and the number of measurements m. Then, on the similar way as for the success rates in Figure 1.3 but only for 200 repetitions, for each pair (m, s) the probability of a successful reconstruction via (4.4) is calculated. As before, it is said that a reconstruction $x^\diamond$ of the original signal $x \in \mathbb{R}^n$ is exact if $\|x - x^\diamond\|_1 \leqslant 10^{-3}$.

The result is shown in Figure 4.1 which presents the so called *phase transition* that occurs in modern high-dimensional data analysis and signal processing [76, 77]. Of course the figure might slightly change for different reconstruction algorithms or measurement matrices used, but the shape mainly remains the same. If one takes $m = 60$ measurements and assumes the signal to be $s = 45$ sparse, it is $\tau = 0.6$ and $\rho = 0.75$. For the coin problem and the recovery algorithm used, Figure 4.1 then states that reconstruction is hopeless since one is then in the white area. Now, only factitiously increasing the size n of the problem is obviously not the useful way to go. If it is wanted to reach the black area one needs to increase both the problem size n and the number m of measurements.

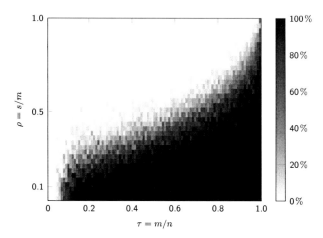

Figure 4.1.: Donoho-Tanner phase transition for the coin problem from Section 1.3 where $n = 100$. Based on 200 repetitions it is shown how high the probability (from 0% to 100%) is for a perfect reconstruction x° of random created forgery vectors x ($\|x - x^\circ\|_1 \leqslant 10^{-3}$) for different number of measurements $m \in \{1, \ldots, n\}$ and sparsity levels $s \in \{1, \ldots, n\}$.

In many applications, the signal $x \in \mathbb{R}^n$ to recover is not sparse in itself, but rather sparse in a basis $\Psi \in \mathbb{R}^{n \times n}$ where the basis vectors are given in the columns. Therefore, $x = \Psi z$ with $\|z\|_0 \leqslant n$ and the minimization problem to solve then reads

$$\operatorname*{argmin}_{z \in \mathbb{R}^n} \|z\|_0 \ \text{ subject to } \ \Phi \Psi z = y. \tag{4.17}$$

It is natural to ask whether the thus constructed matrix $\Phi\Psi$ satisfies the RIP with high probability. In fact one can show the following theorem which proofs that at least orthogonal matrices Ψ inherent this so-called *universality* property [17, 93, 171].

Theorem 4.2.21 (Universality of random matrices [93, Theorem 9.15]). *Let $\Psi \in \mathbb{R}^{n \times n}$ be a (fixed) orthogonal matrix. Suppose that an $m \times n$ random matrix Φ is drawn according to a probability distribution for which the concentration inequality (4.12) holds for all $t \in (0, 1)$ and $x \in \mathbb{R}^n$.*

Then, given $\delta, \varepsilon \in (0,1)$, the restricted isometry constant δ_s of $\Phi\Psi$ satisfies $\delta_s < \zeta$ with probability at least $1 - \varepsilon$ provided

$$m \geq C\zeta^{-2}\left(s\log\left(\frac{n}{m}\right) + \log(\varepsilon^{-1})\right) \tag{4.18}$$

with a constant C only depending on c_0 from the concentration property (4.12).

Proof. Let $\tilde{x} = \Psi x$. Due to the orthogonality of the matrix Ψ and the fact that Φ fulfills the concentration inequality (4.12), it holds

$$\mathbb{P}\left(\left|\|\Phi\Psi x\|_2^2 - \|x\|_2^2\right| \geq t\|x\|_2^2\right) = \mathbb{P}\left(\left|\|\Phi\tilde{x}\|_2^2 - \|\tilde{x}\|_2^2\right| \geq t\|\tilde{x}\|_2^2\right)$$
$$\leq 2\exp(-c_0 t^2 m).$$

The application of Theorem 4.2.17 completes the proof. $\qquad\square$

4.3. Stable ℓ_1 minimization

Instead of solving the non-convex ℓ_0 problem (4.4), it is common to solve the convex relaxation

$$\operatorname*{argmin}_{z \in \mathbb{R}^n} \|z\|_1 \ \text{ subject to } \ \Phi z = y, \tag{4.19}$$

As in (4.17), by taking into account the a priori knowledge that the measured signal is sparse in a basis Ψ, the minimizatin problem reads

$$\operatorname*{argmin}_{z \in \mathbb{R}^n} \|z\|_1 \ \text{ subject to } \ \Phi\Psi z = y. \tag{4.20}$$

In this section it is studied under which conditions a solution to (4.19) still leads to a sparse minimizer. Moreover, by using the RIP it is shown that under some conditions the solutions of (4.19) and its ℓ_0-variant (4.4) are equal.

The following theorem confirms that solutions to (4.19) of (4.4) are still sparse.

Theorem 4.3.1 (Sparseness of ℓ_1 minimizers [93, Theorem 3.1]). *Let $\Phi \in \mathbb{R}^{m \times n}$ with $m \leq n$ be a measurement matrix with columns $\varphi_1, \ldots, \varphi_n$.*

Assuming the uniqueness of a minimizer $x^\circ \in \mathbb{R}^n$ of (4.19), the system $\{\varphi_j \mid j \in \operatorname{supp}(x^\circ)\}$ is linearly independent, and in particular

$$\|x^\circ\|_0 = |\operatorname{supp}(x^\circ)| \leqslant m. \tag{4.21}$$

Proof. By way of contradiction assume that the set $\{\varphi_j \mid j \in S\}$ with $S = \operatorname{supp}(x^\circ)$ is linearly dependent. Therefore, there exists a non-zero vector $w \in \mathbb{R}^n$ supported on S such that $Aw = 0$. For any $t \neq 0$ it then follows

$$\|x^\circ\|_1 < \|x^\circ + tw\|_1 = \sum_{i \in S} |x_i^\circ + tw_i| = \sum_{i \in S} \operatorname{sign}(x_i^\circ + tw_i)(x_i^\circ + tw_i).$$

Now, for small t, namely for $|t| < \min_{i \in S} |x_i^\circ|/\|w\|_\infty$, it holds $|x_i^\circ| > t|w_i|$ for all $i \in S$ and all $t \neq 0$. From this it follows

$$\operatorname{sign}(x_i^\circ + tw_i) = \operatorname{sign}(x_i^\circ),$$

and with that

$$\begin{aligned}
\|x^\circ\|_1 &< \sum_{i \in S} \operatorname{sign}(x_i^\circ)(x_i^\circ + tw_i) \\
&= \sum_{i \in S} \operatorname{sign}(x_i^\circ)x_i^\circ + t \sum_{i \in S} \operatorname{sign}(x_i^\circ)w_i \\
&= \|x^\circ\|_1 + t \sum_{i \in S} \operatorname{sign}(x_i^\circ)w_i.
\end{aligned}$$

Since one can always choose a small $t \neq 0$ such that $t \sum_{i \in S} \operatorname{sign}(x_i^\circ)w_i \leqslant 0$, this is a contradiction. The inequality in (4.21) follows from the fact that there are at most m linearly independent columns possible. $\qquad\square$

In addition to the ℓ_1 minimization problems (4.19) or (4.20), in practice the measurements are usually corrupted with noise w with $\|w\|_2 \leqslant \varepsilon$. In case of additive noise, (4.3) reads

$$y = \Phi x + w. \tag{4.22}$$

In this context, instead of solving (4.19), one aims to solve the following relaxed convex optimization problem, see also Figure 3.2(c), page 32:

$$\operatorname*{argmin}_{z \in \mathbb{R}^n} \|z\|_1 \quad \text{subject to} \quad \|\Phi z - y\|_2 \leqslant \varepsilon. \tag{4.23}$$

Whereas the Lemma 4.2.12 gives only a connection between the RIP and the uniqueness of a s-sparse solution, results that guarantee the robustness of recovery via (4.23) will now be discussed. What is meant by *robust recovery* is the fact that the recovered signal is not too far away from the measured (true) signal with respect to the underlying noise. More precisely the distance between both is bounded by the degree of compressibility and the noise. Moreover, there are several robustness results published over the years, improving the RIP constant of the measurement matrix Φ. In the notion of this thesis, it is stated the one by Candès in [42]. Even though it does not present the newest result, its proof concept is of the kind that is later used in the robust recovery result in Chapter 7.

Theorem 4.3.2 (Robust recovery under noise via RIP [42, Theorem 1.3]). *Let $x \in \mathbb{R}^n$, $x_s := \mathrm{argmin}_{\tilde{x}_s \in \Sigma_s} \|x - \tilde{x}_s\|_2$ and let Φ be a $m \times n$ matrix that satisfies the RIP of order $2s$ with constant $\delta_{2s} < \sqrt{2} - 1$. Let further $x^\diamond$ be the solution to (4.23) and assume that $\|w\|_2 \leqslant \varepsilon$. Then the solution $x^\diamond$ obeys*

$$\|x^\diamond - x\|_2 \leqslant C_0 \frac{1}{\sqrt{s}} \|x - x_s\|_1 + C_1 \varepsilon \qquad (4.24)$$

with some constants C_0 and C_1.

The inequality (4.24) states that if Φ fulfills the required RIP, then the distance from the solution $x^\diamond$ from the optimization problem (4.23) to the exact solution x is bounded by its compressibility level and the noise. It is clear that Theorem 4.3.2 implicitly handles the noiseless recovery case (4.19). If x is exactly s-sparse and there is no noise in the measurements, then the reconstruction is exact. The proof of Theorem 4.3.2 requires the following lemma:

Lemma 4.3.3 ([42, Lemma 2.1]). *Let $\Phi \in \mathbb{R}^{m \times n}$ satisfy the RIP of order $s + s'$ with constant $\delta_{s+s'}$. Let $x, x' \in \mathbb{R}^n$ be vectors supported on disjoint subsets $S, S' \subseteq \{1, \ldots, n\}$ with $|S| \leqslant s$, $|S'| \leqslant s'$. Then*

$$|\langle \Phi x, \Phi x' \rangle| \leqslant \delta_{s+s'} \|x\|_2 \|x'\|_2.$$

Proof. Due to the RIP of Φ it is

$$(1 - \delta_{s+s'}) \|x \pm x'\|_2^2 \leqslant \|\Phi x \pm \Phi x'\|_2^2 \leqslant (1 + \delta_{s+s'}) \|x \pm x'\|_2^2.$$

Now let $\bar{x}$ and $\bar{x}'$ be the unit norm variants of x, x', i.e. $\bar{x} = x/\|x\|_2$ and $\bar{x}' = x'/\|x'\|_2$. Together with the parallelogram identity and using disjoint supports it then follows that

$$
\begin{aligned}
|\langle \Phi\bar{x}, \Phi\bar{x}'\rangle| &= \frac{1}{4}\big|\|\Phi\bar{x} + \Phi\bar{x}'\|_2^2 - \|\Phi\bar{x} - \Phi\bar{x}'\|_2^2\big| \\
&\leqslant \frac{1}{4}\big|(1 + \delta_{s+s'})\|\bar{x} + \bar{x}'\|_2^2 - (1 - \delta_{s+s'})\|\bar{x} - \bar{x}'\|_2^2\big| \\
&= \frac{1}{4}\delta_{s+s'}\big(\|\bar{x} + \bar{x}'\|_2^2 + \|\bar{x} - \bar{x}'\|_2^2\big) \\
&= \frac{1}{4}\delta_{s+s'}\big(2\|\bar{x}\|_2^2 + 2\|\bar{x}'\|_2^2\big) \\
&= \delta_{s+s'}.
\end{aligned}
$$

Multiplying both sides by $\|x\|_2\|x'\|_2$ then concludes the proof. □

Proof of Theorem 4.3.2. At first note that there holds a *tube constraint* of the following form

$$
\|\Phi(x^\circ - x)\|_2 \leqslant \|\Phi x^\circ - y\|_2 + \|y - \Phi x\|_2 \leqslant 2\varepsilon, \tag{4.25}
$$

since x is feasible for problem (4.23). As next, again due to the feasibility of x, it must hold $\|x\|_1 \geqslant \|x^\circ\|_1$. Now set $h \in \mathbb{R}^n$ with $x^\circ = x + h$. Decompose h into a sum of vectors $h_S, h_{S_1}, \ldots, h_{S_r}$,

$$
h = h_S + h_{S_1} + \ldots + h_{S_r} = h_S + h_{S^C},
$$

where $r = \lfloor n/s \rfloor$ and S is the index set of locations of the s largest coefficients of x. S_1 is the index set of the s largest magnitude coefficients of the complement of S, i.e. S^C. S_2 is the set of the next s largest magnitude coefficients with respect to the index set $S^C \backslash S_1$ and so on. In addition, the notation h_S (or analogously h_{S^C}) has to be understood component wise as $h_S^i = h^i$ if the i-th index corresponds to the index set S of x and $h_S^i = 0$ elsewhere.

This setting leads to the identities

$$
\|h_{S_j}\|_2^2 = \sum_{i=1}^{s}|h_{S_j}^i|^2 \leqslant \|h_{S_j}\|_\infty^2 \sum_{i=1}^{s} 1 = s\|h_{S_j}\|_\infty^2 \tag{4.26}
$$

and

$$s\|h_{S_j}\|_\infty = s\max_i |h_{S_j}^i| \leqslant s\frac{1}{s}\|h_{S_{j-1}}\|_1 = \|h_{S_{j-1}}\|_1. \qquad (4.27)$$

The inequality in (4.27) holds by construction of the index sets, as they for all $j \geqslant 2$ and all k imply

$$|h_{S_j}^k| \leqslant \frac{1}{s}\sum_i |h_{S_{j-1}}^i| = \frac{1}{s}\|h_{S_{j-1}}\|_1,$$

meaning that each average magnitude coefficient of h based on the previous index set S_{j-1} is greater or equal to any other coefficient of h based on the set S_j. Both, (4.26) and (4.27) together yield

$$\|h_{S_j}\|_2 \leqslant \sqrt{s}\|h_{S_j}\|_\infty \leqslant \frac{1}{\sqrt{s}}\|h_{S_{j-1}}\|_1. \qquad (4.28)$$

Due to the above mentioned feasibility of x one can observe that

$$\begin{aligned}
\|x\|_1 &\geqslant \|x + h\|_1 \\
&= \sum_{i\in S}|x_i + h_i| + \sum_{i\in S^C}|x_i + h_i| \\
&\geqslant \sum_{i\in S}\left(|x_i| - |h_i|\right) + \sum_{i\in S^C}\left(|h_i| - |x_i|\right) \\
&= \|x_S\|_1 - \|x_{S^C}\|_1 - \|h_S\|_1 + \|h_{S^C}\|_1,
\end{aligned} \qquad (4.29)$$

which leads to a *cone constraint*

$$\|h_{S^C}\|_1 \leqslant \|h_S\|_1 + 2\|x_{S^C}\|_1. \qquad (4.30)$$

Note that by definition $\|x_{S^C}\|_1 = \|x - x_s\|_1$. Now, because of (4.28), (4.30) and

$$\sum_{j\geqslant 2}\|h_{S_j}\|_2 \leqslant \frac{1}{\sqrt{s}}\sum_{j\geqslant 1}\|h_{S_j}\|_1 = \frac{1}{\sqrt{s}}\|h_{S^C}\|_1, \qquad (4.31)$$

it holds

$$\begin{aligned}
\|h_{(S\cup S_1)^C}\|_2 = \Big\|\sum_{j\geqslant 2}h_{S_j}\Big\|_2 &\leqslant \sum_{j\geqslant 2}\|h_{S_j}\|_2 \leqslant \frac{1}{\sqrt{s}}\|h_{S^C}\|_1 \\
&\leqslant \frac{1}{\sqrt{s}}\|h_S\|_1 + 2\frac{1}{\sqrt{s}}\|x_{S^C}\|_1 \\
&\leqslant \|h_S\|_2 + 2\eta,
\end{aligned} \qquad (4.32)$$

where $\eta := \frac{1}{\sqrt{s}}\|x - x_s\|_1$. The last inequality follows from the Cauchy-Schwarz inequality as $\|h_S\|_1 = |\langle h_S, \mathbb{1}\rangle| \leqslant \|h_S\|_2 \|\mathbb{1}\|_2 = \sqrt{s}\|h_S\|_2$. Then, because of $\Phi h_{S \cup S_1} = \Phi h - \sum_{j \geqslant 2} \Phi h_{S_j}$, it follows that

$$\|\Phi h_{S \cup S_1}\|_2^2 = \langle \Phi h_{S \cup S_1}, \Phi h\rangle - \left\langle \Phi h_{S \cup S_1}, \sum_{j \geqslant 2} \Phi h_{S_j} \right\rangle. \tag{4.33}$$

Applying the Cauchy-Schwarz inequality, the tube constraint (4.25) and the restricted isometry property of Φ, it is

$$|\langle \Phi h_{S \cup S_1}, \Phi h\rangle| \leqslant \|\Phi h_{S \cup S_1}\|_2 \|\Phi h\|_2 \leqslant 2\varepsilon\sqrt{1 + \delta_{2s}}\|h_{S \cup S_1}\|_2.$$

With (4.33), Lemma 4.3.3, the inequality[3]

$$\|h_S\|_2 + \|h_{S_1}\|_2 \leqslant \sqrt{2}\|h_{S \cup S_1}\|_2,$$

and because of

$$\left| \left\langle \Phi h_{S \cup S_1}, \sum_{j \geqslant 2} \Phi h_{S_j} \right\rangle \right| = \left| \left\langle \Phi h_S + \Phi h_{S_1}, \sum_{j \geqslant 2} \Phi h_{S_j} \right\rangle \right|$$

$$\leqslant \left| \left\langle \Phi h_S, \sum_{j \geqslant 2} \Phi h_{S_j} \right\rangle \right| + \left| \left\langle \Phi h_{S_1}, \sum_{j \geqslant 2} \Phi h_{S_j} \right\rangle \right|$$

$$\leqslant \sum_{j \geqslant 2} \left(|\langle \Phi h_S, \Phi h_{S_j}\rangle| + |\langle \Phi h_{S_1}, \Phi h_{S_j}\rangle| \right)$$

$$\leqslant \sum_{j \geqslant 2} \delta_{2s}\|h_{S_j}\|_2 \left(\|h_S\|_2 + \|h_{S_1}\|_2 \right)$$

$$\leqslant \sqrt{2}\delta_{2s}\|h_{S \cup S_1}\|_2 \sum_{j \geqslant 2} \|h_{S_j}\|_2,$$

it follows

$$(1 - \delta_{2s})\|h_{S \cup S_1}\|_2^2 \leqslant \|\Phi h_{S \cup S_1}\|_2^2$$

$$\leqslant 2\varepsilon\sqrt{1 + \delta_{2s}}\|h_{S \cup S_1}\|_2 + \left| \left\langle \Phi h_{S \cup S_1}, \sum_{j \geqslant 2} \Phi h_{S_j} \right\rangle \right|$$

$$\leqslant \|h_{S \cup S_1}\|_2 \left(2\varepsilon\sqrt{1 + \delta_{2s}} + \sqrt{2}\delta_{2s} \sum_{j \geqslant 2} \|h_{S_j}\|_2 \right).$$

---

[3]By use of Lemma 2.1.3, page 17, one shows $x + y - \frac{1}{2}(\sqrt{x} + \sqrt{y})^2 \geqslant 0$ for all $x, y \geqslant 0$. The inequality then follows by setting $x = \|h_S\|_2^2$, $y = \|h_{S_1}\|_2^2$ and using the fact that S and S_1 are disjoint.

Together with (4.31) this yields

$$\|h_{S \cup S_1}\|_2 \leqslant \alpha_1 \varepsilon + \alpha_2 \frac{1}{\sqrt{s}} \|h_{S^C}\|_1, \tag{4.34}$$

where $\alpha_1 = \frac{2\sqrt{1+\delta_{2s}}}{1-\delta_{2s}}$ and $\alpha_2 = \frac{\sqrt{2}\delta_{2s}}{1-\delta_{2s}}$. Applying the cone constraint (4.30), the Cauchy-Schwarz inequality, the fact that $\|h_S\|_2 \leqslant \|h_{S \cup S_1}\|_2$ as well as the identity $\frac{1}{\sqrt{s}}\|h_{S^C}\|_1 \leqslant \|h_S\|_2 + 2\eta$ from (4.32), it follows

$$\begin{aligned} \|h_{S \cup S_1}\|_2 &\leqslant \alpha_1 \varepsilon + \alpha_2 \|h_{S \cup S_1}\|_2 + 2\alpha_2 \eta \\ \Longleftrightarrow \quad \|h_{S \cup S_1}\|_2 &\leqslant (1 - \alpha_2)^{-1}(\alpha_1 \varepsilon + 2\alpha_2 \eta), \end{aligned} \tag{4.35}$$

Together with (4.32) one concludes

$$\|h\|_2 \leqslant \|h_{S \cup S_1}\|_2 + \|h_{(S \cup S_1)^C}\|_2 \leqslant 2(1-\alpha_2)^{-1}(\alpha_1 \varepsilon + (1+\alpha_2)\eta),$$

which ends the proof. $\qquad\square$

Even though this proof looks rather technical, it is at most basic linear algebra. The condition on the restricted isometry constant to be $\delta_{2s} < \sqrt{2} - 1$ is essential in (4.35) where it is indispensable to have $1 - \alpha_2 > 0$.

As it has been discussed right after Theorem 4.2.19, the larger δ_{2s}, the more it is guaranteed that a (Gaussian) measurement matrix has the desired restricted isometry constant. In [48], Candès and Tao have shown that $\delta_{2s} < 1$ is sufficient for the existence of an algorithm to recover all s-sparse vectors x from the measurements $y = Ax$, see also Theorem 4.2.12. Candès published in [42] the condition $\delta_{2s} < \sqrt{2} - 1 \approx 0.4142$, as presented in Theorem 4.3.2. To mention only two more: Foucart and Lai [92] were able to show that $\delta_{2s} < 2/(3 + \sqrt{2}) \approx 0.4531$, and Rauhut *et al.* [93] that $\delta_{2s} < 0.6246$ is sufficient. For more restricted isometry constants and their relationships refer to [41] and [93] and the references therein.

Remark 4.3.4. Geometrically, the *tube constraint* (4.25) means that the solution $x^\diamond$ of (4.23) lies within a cylinder around the plane Φx, see also Figure 3.2(c) on page 32 where both are visualized as a gray area and a dashed line, respectively. Equation (4.30) is known as a *cone constraint*, meaning that it expresses the geometrical fact that h must lie in the ℓ_1-cone at x, cf. Figure 3.2(c). This can also be seen by the first inequality

$\|x\|_1 \geqslant \|x + h\|_1$ in (4.29). For this to hold, h is not allowed to lie out of the ℓ_1-cone descent. In any case, both the tube and the cone constraint ensures that the solution x° lies within the intersection of the cylinder and the ℓ_1-ball.

Now the field of robust recovery via ℓ_1 minimization will be left behind. Instead, stable image reconstructions from CS measurements via total variation minimization will be studied. In light of the definition of the discrete TV norm (2.3), this requires to have stable or robust gradient reconstructions from only few gradient measurements.

4.4. Stable total variation minimization

Stable image reconstruction from linear CS measurements using total variation minimization has only recently been studied in [118, 150, 151]. Let $\mathcal{M}$ be a measurement operator $\mathcal{M} : \mathbb{R}^{n \times n} \to \mathbb{R}^m$. The convex optimization problem for the reconstruction of an image $X \in \mathbb{R}^{n \times n}$ from compressed measurements $y = \mathcal{M}X \in \mathbb{R}^m$ then reads

$$\operatorname*{argmin}_{Z \in \mathbb{R}^{n \times n}} \|Z\|_{TV} \quad \text{subject to} \quad \|\mathcal{M}Z - y\|_2 \leqslant \varepsilon. \qquad (4.36)$$

The works [151] and [150] by Needell and Ward are the first two concerning this topic. In [151] a proof is given for stable image recovery for one image via (4.36). The paper [150] then easily adds this result to any hyperspectral image. The work [118] by Krahmer and Ward extends the previous mentioned articles even further by, for example, formulating bounds on the number of Fourier measurement needed for stable image recovery.

With regard to this thesis, the main result from [151] is stated as well as its proof as its concept is fundamental for the later stability results in Chapter 7.

For the next the Definition 4.2.11 of the RIP is extended for linear operators $\Phi : \mathbb{R}^{n_x \times n_y} \to \mathbb{R}^m$.

Definition 4.4.1. A linear operator $\Phi : \mathbb{R}^{n_x \times n_y} \to \mathbb{R}^m$ has the *restricted isometry property* (RIP) of order s and level $\delta \in (0, 1)$ if

$$(1 - \delta)\|X\|_F^2 \leqslant \|\Phi X\|_2^2 \leqslant (1 + \delta)\|X\|_F^2$$

for all s-sparse matrices $X \in \mathbb{R}^{n_x \times n_y}$.

The following Proposition 4.4.2 uses the concepts of a tube and a cone constraint as discussed in the Remark 4.3.4. It lays the foundation for the proof of the main Theorem 4.4.7 that will be presented thereafter.

Proposition 4.4.2 ([151, Proposition 3]). *Fix parameters $\gamma \geqslant 1$, $\delta < 1/3$ and let $k \in \mathbb{N}$, $C > 0$ and $\varepsilon, \sigma \geqslant 0$. Suppose that $\mathcal{A} : \mathbb{R}^{n \times n} \to \mathbb{R}^m$ satisfies the RIP of order $5k\gamma^2$ and level δ, and suppose that the image $D \in \mathbb{R}^{n \times n}$ satisfies a tube constraint*

$$\|\mathcal{A}D\|_2 \leqslant C\varepsilon. \tag{4.37}$$

Suppose further that for a subset S of cardinality $|S| \leqslant k$, D satisfies the cone constraint

$$\|D_{S^C}\|_1 \leqslant \gamma\|D_S\|_1 + \sigma. \tag{4.38}$$

Then

$$\|D\|_F \lesssim \frac{\sigma}{\gamma\sqrt{k}} + \varepsilon,$$

and

$$\|D\|_1 \lesssim \sigma + \gamma\sqrt{k}\varepsilon.$$

The proof mainly follows the same principle as for the proof of Theorem 4.3.2, see also [42] or [47].

Proof. Let $s = k\gamma^2$ and let $S \subseteq \{1, \ldots, n^2\}$ be the support set of the best s-term approximation of D. Decompose $D = D_S + D_{S^C}$ where

$$D_{S^C} = D_{S_1} + D_{S_2} + \ldots + D_{S_r}$$

and $r = \lfloor \frac{n^2}{4s} \rfloor$ as follows. Similar to the proof of Theorem 4.3.2, D_{S_1} consists of the $4s$ largest magnitude coefficients of D over S^C, D_{S_2} consists of the $4s$ largest magnitude coefficients of D over $S^C \backslash S_1$ and so on. Also, by construction it holds that the average magnitude of each nonzero coefficient $D_{S_{j-1}}^{k,l}$ of $D_{S_{j-1}}$ is larger than the magnitude of each of the nonzero coefficients $D_{S_j}^{t,p}$ of D_{S_j}, i.e. for all $j = 2, 3, \ldots, r$ it holds

$$\frac{1}{4s} \sum_{k,l}^{4s} |D_{S_{j-1}}^{k,l}| \geqslant |D_{S_j}^{t,p}|.$$

Squaring both sides and summing over all $4s$ nonzero components of D_{S_j} then leads to

$$\|D_{S_j}\|_F \leqslant \frac{\|D_{S_{j-1}}\|_1}{2\sqrt{s}}.$$

Together with the cone constraint (4.38) this gives

$$\begin{aligned}
\sum_{j=2}^{r} \|D_{S_j}\|_F &\leqslant \frac{1}{2\gamma\sqrt{k}} \|D_{S^C}\|_1 \\
&\leqslant \frac{1}{2\sqrt{k}} \|D_S\|_1 + \frac{1}{2\gamma\sqrt{k}}\sigma \qquad (4.39) \\
&\leqslant \frac{1}{2} \|D_S\|_F + \frac{1}{2\gamma\sqrt{k}}\sigma.
\end{aligned}$$

Combining (4.39) with the RIP for $\mathcal{A}$ as well as the tube constraint (4.37) yields

$$\begin{aligned}
C\varepsilon &\geqslant \|\mathcal{A}D\|_2 \\
&\geqslant \|\mathcal{A}(D_S + D_{S_1})\|_2 - \sum_{j=2}^{r} \|\mathcal{A}D_{S_j}\|_2 \\
&\geqslant \sqrt{1-\delta}\|D_S + D_{S_1}\|_F - \sqrt{1+\delta}\sum_{j=2}^{r} \|D_{S_j}\|_F \\
&\geqslant \sqrt{1-\delta}\|D_S + D_{S_1}\|_F - \sqrt{1+\delta}\left(\frac{1}{2}\|D_S\|_F + \frac{1}{2\gamma\sqrt{k}}\sigma\right) \\
&\geqslant \left(\sqrt{1-\delta} - \frac{\sqrt{1+\delta}}{2}\right)\|D_S + D_{S_1}\|_F - \sqrt{1+\delta}\frac{1}{2\gamma\sqrt{k}}\sigma.
\end{aligned}$$

Since $\delta < 1/3$, it follows[4]

$$\|D_S + D_{S_1}\|_F \leqslant 5C\varepsilon + \frac{3\sigma}{\gamma\sqrt{k}}. \tag{4.40}$$

Due to the inequality (4.39) for the other components of the complement D_{S_j},

$$\left\|\sum_{j=2}^{r} D_{S_j}\right\|_F \leqslant \sum_{j=2}^{r} \|D_{S_j}\|_F \leqslant \frac{1}{2}\|D_S + D_{S_1}\|_F + \frac{1}{2\gamma\sqrt{k}}\sigma,$$

one concludes

$$\|D\|_F \leqslant 8C\varepsilon + \frac{5\sigma}{\gamma\sqrt{k}}.$$

The last inequality to show follows from the cone constraint (4.38), Lemma 2.1.1 as well as (4.40),

$$\begin{aligned}
\|D\|_1 &= \|D_S\|_1 + \|D_{S^C}\|_1 \\
&\leqslant (\gamma + 1)\|D_S\|_1 + \sigma \\
&\leqslant 2\gamma\sqrt{s}\|D_S\|_F + \sigma \\
&\leqslant 2\gamma\sqrt{s}\left(5C\varepsilon + \frac{3\sigma}{\gamma\sqrt{k}}\right) + \sigma \\
&\leqslant 6\gamma\sigma + 10C\gamma^2\sqrt{k}\varepsilon,
\end{aligned}$$

which concludes the proof. □

The proof of the following Theorem 4.4.7 describing stable image recovery via total variation from compressed measurements makes use of compressed measurements on the gradient. The discrete total variation norm is defined as the 1-norm on the gradient. The proof concept requires to have measurements on the image gradients as well. For this reason, the following notation from [151] will be used.

---

[4]Similar as in Theorem 4.3.2, the assumption on the restricted isometry constant δ arises from the condition to have the term $\sqrt{1-\delta} - \frac{\sqrt{1+\delta}}{2}$ positive for further calculation.

For a matrix Φ, Φ_0 and Φ^0 denote the matrices which arise from Φ by concatenating a row of zeros at the bottom or on top, respectively. More formally, for a matrix $\Phi \in \mathbb{R}^{(n_x-1) \times n_y}$ the augmented matrix $\Phi^0 \in \mathbb{R}^{n_x \times n_y}$ is component-wise defined as

$$(\Phi^0)_{i,j} = \begin{cases} 0, & i = 1 \\ \Phi_{i-1,j}, & 2 \leqslant i \leqslant n_x. \end{cases} \tag{4.41}$$

The augmented matrix $\Phi_0 \in \mathbb{R}^{n_x \times n_y}$ is defined accordingly.

The following lemma establishes a relation between compressed sensing measurements of directional gradients and these padded matrices.

Lemma 4.4.3. *Let $X \in \mathbb{R}^{n_x \times n_y}$, $\Phi \in \mathbb{R}^{(n_x-1) \times n_y}$ and $\Psi \in \mathbb{R}^{(n_y-1) \times n_x}$. Then*

$$\langle \Phi, X_x \rangle = \langle \Phi^0, X \rangle - \langle \Phi_0, X \rangle$$

and

$$\langle \Psi, X_y^T \rangle = \langle \Psi^0, X^T \rangle - \langle \Psi_0, X^T \rangle,$$

where the scalar product $\langle \cdot, \cdot \rangle$ is defined as in (2.1), page 15, and the derivatives $X_x \in \mathbb{R}^{(n_x-1) \times n_y}$ and $X_y \in \mathbb{R}^{n_x \times (n_y-1)}$ are given as described in (2.4), page 16.

Remark 4.4.4. Lemma 4.4.3 is a relative abstract form to describe what is later on meant by *gradient measurements*. The expressions $\langle \Phi, X_x \rangle$ and $\langle \Psi, X_y^T \rangle$ describe the multiplication of the derivatives X_x and X_y using the matrices Φ and Ψ.

Proof of Lemma 4.4.3. Using the definitions of the directional derivatives and the inner product, simple algebraic manipulations lead to

$$\langle \Phi, X_x \rangle = \sum_{\substack{1 \leqslant i \leqslant n_x - 1 \\ 1 \leqslant j \leqslant n_y}} \Phi_{i,j} (X_x)_{i,j}$$

$$= \sum_{\substack{1 \leqslant i \leqslant n_x - 1 \\ 1 \leqslant j \leqslant n_y}} \Phi_{i,j} (X_{i+1,j} - X_{i,j})$$

$$= \sum_{\substack{1 \leqslant i \leqslant n_x \\ 1 \leqslant j \leqslant n_y}} \left(\Phi_{i,j}^0 X_{i+1,j} - \Phi_{0,i,j} X_{i,j} \right)$$

$$= \langle \Phi^0, X \rangle - \langle \Phi_0, X \rangle.$$

The other equality follows similarly. □

For the rest of this section it will be $n_x = n_y =: n$, i.e. X will be a squared image $X \in \mathbb{R}^{n \times n}$.

The robustness Theorem 4.4.7 will also make use of linear operators of the kind

$$\mathcal{A} : \mathbb{R}^{(n-1) \times n} \to \mathbb{R}^{m_1} \quad \text{and} \quad \mathcal{A}' : \mathbb{R}^{(n-1) \times n} \to \mathbb{R}^{m_1}. \tag{4.42}$$

Moreover, in light of Lemma 4.4.3, concatenated versions of (4.42) will be used. Their formal definition is given next.

Remark 4.4.5. In the following, the linear operator $\mathcal{A} : \mathbb{R}^{(n-1) \times n} \to \mathbb{R}^m$ describes m measurements of a matrix $Z \in \mathbb{R}^{(n-1) \times n}$. They are formulated as $j = 1, \dots, m$ inner products

$$(\mathcal{A}Z)_j = \langle A_j, Z \rangle$$

with measurement matrices $A_j \in \mathbb{R}^{(n-1) \times n}$. For a matrix $X \in \mathbb{R}^{n \times n}$, the augmented variants

$$\mathcal{A}^0 : \mathbb{R}^{n \times n} \to \mathbb{R}^m \quad \text{and} \quad \mathcal{A}_0 : \mathbb{R}^{n \times n} \to \mathbb{R}^m \tag{4.43}$$

are for $j = 1, \dots, m$ defined accordingly as

$$(\mathcal{A}^0 X)_j = \langle (A^0)_j, X \rangle \quad \text{and} \quad (\mathcal{A}_0 X)_j = \langle (A_0)_j, X \rangle, \tag{4.44}$$

where $(A^0)_j \in \mathbb{R}^{n \times n}$ and $(A_0)_j \in \mathbb{R}^{n \times n}$ are the padded variants of the above mentioned measurement matrices A_j.

Remark 4.4.6. As mentioned above, each of them describes m_1 taken measurements of a matrix $Z \in \mathbb{R}^{(n-1) \times n}$. In light of Remark 4.4.5, let $A_1, A_2, \dots, A_{m_1}$ and $A'_1, A'_2, \dots, A'_{m_1}$ be the m_1 components of the horizontal and vertical gradient measurements, i.e.

$$(\mathcal{A}Z)_j = \langle A_j, Z \rangle \quad (\mathcal{A}'Z)_j = \langle A'_j, Z \rangle.$$

In light of gradient measurements, the matrix Z will in Theorem 4.4.7 be replaced by derivatives of a given matrix. Hence, they are both matrices of size $(n-1) \times n$. $\mathcal{A}$ will be an operator which takes m_1 measurements of the derivative of a matrix in x-direction. $\mathcal{A}'$ does the same with respect to the derivative in y-direction.

Together with Lemma 4.4.3 and the definitions given in (4.43) and (4.44), full gradient measurements of a given image $X \in \mathbb{R}^{n \times n}$ are described via

$$\left(\mathcal{A}^0 X, \mathcal{A}_0 X, \mathcal{A}'^0 X^T, \mathcal{A}'_0 X^T \right).$$

Theorem 4.4.7 (Stable recovery by TV using RIP measurements [151, Theorem 5]). *Let $n = 2^N$ be a power of two. Let $\mathcal{A} : \mathbb{R}^{(n-1) \times n} \to \mathbb{R}^{m_1}$ and $\mathcal{A}' : \mathbb{R}^{(n-1) \times n} \to \mathbb{R}^{m_1}$ be such that the concatenated operator $[\mathcal{A} \; \mathcal{A}'] : \mathbb{R}^{2(n-1) \times n} \to \mathbb{R}^{2m_1}$ has the RIP of order $5s$ and level $\delta < 1/3$. Let the linear operator $\Phi : \mathbb{R}^{n \times n} \to \mathbb{R}^{m_2}$ has the RIP of order $2s$ and level $\delta < 1$. Let $m = 4m_1 + m_2$, and consider the linear operator $\mathcal{M} : \mathbb{R}^{n \times n} \to \mathbb{R}^m$ with components*

$$\mathcal{M}(X) = \left(\mathcal{A}^0 X, \mathcal{A}_0 X, \mathcal{A}'^0 X^T, \mathcal{A}'_0 X^T, \Phi X \right). \tag{4.45}$$

Let $X \in \mathbb{R}^{n \times n}$ be an image with discrete gradient ∇X. If noisy measurements $y = \mathcal{M}X + \xi$ are observed with noise level $\|\xi\|_2 < \varepsilon$, then

$$X^\diamond = \underset{Z \in \mathbb{R}^{n \times n}}{\operatorname{argmin}} \|Z\|_{TV} \qquad \text{such that} \qquad \|\mathcal{M}Z - y\|_2 \leqslant \varepsilon \tag{4.46}$$

satisfies

$$\|\nabla X - \nabla X^\diamond\|_F \lesssim \frac{\|\nabla X - (\nabla X)_s\|_1}{\sqrt{s}} + \varepsilon \tag{4.47}$$

and

$$\|X - X^\diamond\|_{TV} \lesssim \|\nabla X - (\nabla X)_s\|_1 + \sqrt{s}\varepsilon. \tag{4.48}$$

It is worth providing a few remarks on the setting of the Theorem 4.4.7 as well as on the Proposition 4.4.2, as their style will appear later on in Chapter 7 where stable recovery for compressed sensed MALDI data is shown.

Remark 4.4.8. The linear operator $\mathcal{M}$ in (4.45) consists of two parts. The first is given by compressed gradient measurements modelled by the concatenated versions $\mathcal{A}^0$, $\mathcal{A}_0$, $\mathcal{A}'^0$ and $\mathcal{A}'_0$ of the linear operators $\mathcal{A}$ and $\mathcal{A}'$. The second part is given by the the linear measurement operator $\Phi : \mathbb{R}^{n \times n} \to \mathbb{R}^{m_2}$ and is principally the main CS operation on the image $X \in \mathbb{R}^{n \times n}$. Its RIP order of being $2s$ with constant $\delta < 1$ is not directly needed for the proof, but rather a hedge of unique reconstruction in the sense of Lemma 4.2.12.

Remark 4.4.9. The condition to have images of side-length $n = 2^N$ for some integer N is not a restriction, as one can horizontally and vertically reflect the image to have a size $2n \times 2n$ with the same TV norm up to a factor 4.

Remark 4.4.10. The appearing order $5s$ for the concatenated operator $[\mathcal{A} \; \mathcal{A}']$ comes from the proof of Proposition 4.4.2, where one decomposes the image D at some point into $D_S + D_{S_1}$, i.e. into a sum of "images" of s and $4s$ sparsity, respectively.

Proof of Theorem 4.4.7. Since the concatenated linear operator $[\mathcal{A} \; \mathcal{A}']$ satisfies the RIP, for Proposition 4.4.2 to apply it suffices to show that $\nabla(X - X^\diamond)$, concatenated as a vector, fulfills the tube and cone constraints.

For this let $D = X - X^\diamond \in \mathbb{R}^{n \times n}$ and $L = (D_x, D_y^T)^T \in \mathbb{R}^{2(n-1) \times n}$ where $D_x \in \mathbb{R}^{(n-1) \times n}$ and $D_y \in \mathbb{R}^{n \times (n-1)}$ are defined as noted in the Preliminaries 2. Let P denote the mapping of indices which maps each nonzero entry of $\nabla D \in \mathbb{R}^{n \times n \times 2}$ to its corresponding index in L. Since L has the same norm as ∇D, i.e. $\|L\|_F = \|\nabla D\|_F$ and $\|L\|_1 = \|\nabla D\|_1$, it suffices to show that L satisfies the tube and cone constraint.

Cone constraint: Let S be the support set of the s largest entries of ∇X. Since X is a feasible solution for (4.46), on a similar way as in the proof of Theorem 4.3.2, there holds

$$\|(\nabla X)_S\|_1 - \|(\nabla D)_S\|_1 - \|(\nabla X)_{S^C}\|_1 + \|(\nabla D)_{S^C}\|_1$$
$$\leqslant \|(\nabla X)_S - (\nabla D)_S\|_1 + \|(\nabla X)_{S^C} - (\nabla D)_{S^C}\|_1$$
$$= \|\nabla X^\diamond\|_1$$
$$\leqslant \|\nabla X\|_1$$
$$= \|(\nabla X)_S\|_1 + \|(\nabla X)_{S^C}\|_1.$$

A reordering provides

$$\|(\nabla D)_{S^C}\|_1 \leqslant \|(\nabla D)_S\|_1 + \|\nabla X - (\nabla X)_s\|_1.$$

Transferring to L and using the fact that $|P(S)| \leqslant |S| \leqslant s$, L satisfies the cone constraint

$$\|L_{P(S)^C}\|_1 \leqslant \|L_{P(S)}\|_1 + 2\|\nabla X - (\nabla X)_s\|_1.$$

Tube constraint: At first note that D fulfills a tube constraint as

$$\|\mathcal{M}D\|_2 \leqslant \|\mathcal{M}X - y\|_2 + \|\mathcal{M}X^\diamond - y\|_2 \leqslant 2\varepsilon. \tag{4.49}$$

Furthermore, applying Lemma 4.4.3 and the parallelogram identity, it follows

$$\begin{aligned}
|\langle A_j, D_x \rangle|^2 &= |\langle A_j^0, D \rangle - \langle A_{j,0}, D \rangle|^2 \\
&\leqslant 2|\langle A_j^0, D \rangle|^2 + 2|\langle A_{j,0}, D \rangle|^2
\end{aligned}$$

and

$$\begin{aligned}
|\langle A'_j, D_y^T \rangle|^2 &= |\langle A'^0_j, D^T \rangle - \langle A'_{j,0}, D^T \rangle|^2 \\
&\leqslant 2|\langle A'^0_j, D^T \rangle|^2 + 2|\langle A'_{j,0}, D^T \rangle|^2.
\end{aligned}$$

Therefore, together with (4.49), L also satisfies a tube constraint of the form

$$\|[\mathcal{A}\ \mathcal{A}'](L)\|_2^2 = \sum_{j=1}^m |\langle A_j, D_x \rangle|^2 + |\langle A'_j, D_y^T \rangle|^2 \leqslant 2\|\mathcal{M}D\|_2^2 \leqslant 8\varepsilon^2.$$

An application of Proposition 4.4.2 then completes the proof. $\qquad\square$

Remark 4.4.11. The authors in [151] have not only shown the robustness results (4.47) and (4.48) in Theorem 4.4.7, but also a stability guarantee which is as follows

$$\|X - X^\diamond\|_F \lesssim \log\left(\frac{n^2}{s}\right)\left(\frac{\|\nabla X - (\nabla X)_s\|_1}{\sqrt{s}} + \varepsilon\right).$$

For this to prove they assume to have compressibility of an image in the wavelet domain. In particular, the authors make use of bivariate Haar coefficients and the fact that their rate of decay can be bounded by the total variation.

Moreover, the authors note that they believe that the additional $4m_1$ gradient measurements are not necessary and only artifacts from the proof.

4.5. Coherent and redundant dictionaries

As it was mentioned in Section 4.2.2, orthogonal matrices Ψ have the property to be *universal* in the sense that if the sensing matrix Φ fulfills the RIP of certain order $\delta_s < 1$, so does the multiplication $\Phi\Psi$ with overwhelming probability, see Theorem 4.2.21. The robustness result in Theorem 4.24 for minimization problems of the form

$$\operatorname*{argmin}_{z \in \mathbb{R}^n} \|z\|_1 \text{ subject to } \Phi\Psi z = y,$$

or in the noisy case

$$\operatorname*{argmin}_{z \in \mathbb{R}^n} \|z\|_1 \text{ subject to } \|\Phi\Psi z - y\|_2 \leqslant \varepsilon, \tag{4.50}$$

holds as long as one can ensure that the basis Ψ in which the signal is supposed to be sparse in is orthogonal.

In this section an extension of Definition 4.2.11 of the RIP as well as a correspondent robustness result will be presented which were both first published in [43]. The extension makes use of the notion of a dictionary as introduced in Definition 3.3.1, page 28. With this notion in hand note the following definition of the *dictionary restricted isometry property* which is a natural extension to the usual RIP.

Definition 4.5.1 (Dictionary restricted isometry property). A matrix $\Phi \in \mathbb{R}^{m \times n}$ has the *dictionary restricted isometry property* (D-RIP) of order s and level $\delta^* \in (0, 1)$, adapted to a dictionary $D \in \mathbb{R}^{n \times p}$, if for all s-sparse $v \in \mathbb{R}^p$ it holds

$$(1 - \delta^*)\|Dv\|_2^2 \leqslant \|\Phi Dv\|_2^2 \leqslant (1 + \delta^*)\|Dv\|_2^2. \tag{4.51}$$

It can be shown that Gaussian matrices or other random matrices satisfy the D-RIP with overwhelming probability, provided that $m \gtrsim s \log(p/s)$, see e.g. [43] and [160]. Equipped with this property, Candès *et al.* have proven the following Theorem 4.5.2.

Theorem 4.5.2 (Stable ℓ_1 recovery via D-RIP [43, Theorem 1.4]). *Let $D \in \mathbb{C}^{n \times p}$ be a (potential complex-valued) dictionary such that there is a constant $C > 0$ so that $D^*D = \frac{1}{C}\,\mathrm{Id}$. This implies that the columns of $\frac{1}{\sqrt{C}}D$ are orthonormal. Let Φ be a measurement matrix satisfying the D-RIP (4.51) with $\delta_{2s} < 0.08$. Then the solution x° to the minimization problem*

$$\operatorname*{argmin}_{\hat{x} \in \mathbb{R}^n} \|D^*\hat{x}\|_1 \quad subject\ to \quad \|\Phi\hat{x} - y\|_2 \leqslant \varepsilon \qquad (4.52)$$

satisfies

$$\|x - x^\circ\|_2 \leqslant C_0\varepsilon + C_1 \frac{\|D^*x - (D^*x)_s\|_1}{\sqrt{s}}, \qquad (4.53)$$

where the constants C_0 and C_1 may only depend on δ_{2s}, $(\cdot)^*$ denotes the conjugate transpose of its expression and $(D^*x)_s$ is the vector consisting of the largest s entries of D^*x in magnitude.

Proof. The proof follows the identical principles as in the proof of Theorem 4.24: One first decomposes the difference $D^*h = D^*(x - x^\circ)$ via subsets of columns of D into disjunct and descent s sparse parts or columns, i.e. $D^*h = D_S^*h + D_{S^C}^*h$ where S is such that $|S| \leqslant s$. One then proves that there holds a tube constraint for the vector Φh and a cone constraint for the vector D^*h. Incorporating the D-RIP then concludes the proof. For further details refer to the proof in [43]. $\qquad\square$

It needs to be mentioned that the so-called ℓ_1-*analysis* problem (4.52) is different from the so-called ℓ_1-*synthesis* approach in (4.50). Where the first seeks for a solution in the signal domain, so does the latter in the coefficient domain. Both formulations are known to be close, but not always equivalent, especially when overcomplete dictionaries come into play. The reader is referred to [85] and the references therein for a detailed comparison of both minimization problems. However, it is shown in [124] that also ℓ_1 recovery using the synthesis formulation (4.50) is robust. The authors from [124] also uses the concept of the D-RIP and arrive at an error bound that is very similar to that in (4.53). At last it is mentioned that Liu *et al.* improved in [125] the D-RIP constant in Theorem 4.5.2 from $\delta_{2s} = 0.08$ to $\delta_{2s} = 0.20$ using a shifting inequality [39].

4.6. Asymmetric restricted isometry property

As it was first independently mentioned in the papers by Blanchard *et al.* in [26] and Foucart *et al.* [92], the symmetrical appearance of the RIP constant in Definition 4.2.11,

$$(1 - \delta)\|x\|_2^2 \leqslant \|\Phi x\|_2^2 \leqslant (1 + \delta)\|x\|_2^2, \tag{4.54}$$

is unnecessarily restrictive. Within this section this observation is briefly motivated and described. The goal is to present an alternative asymmetric definition of the RIP.

First note that the RIP (4.54) holds for all $x \in \Sigma_s$ and can be equivalently written as

$$|\langle (\Phi_S^T \Phi_S - \mathrm{Id})x, x \rangle| \leqslant \delta\|x\|_2 \tag{4.55}$$

for all $S \subset \{1, \ldots, n\}, |S| \leqslant s$. By taking the supremum over all unit norm vectors, it follows

$$\sup_{\|x\|_2=1} |\langle (\Phi_S^T \Phi_S - \mathrm{Id})x, x \rangle| \leqslant \delta.$$

Now, since $\Phi_S^T \Phi_S - \mathrm{Id}$ is self-adjoint it can be diagonalized as $\Phi_S^T \Phi_S - \mathrm{Id} = U^T D U$ with an orthogonal matrix U and diagonal D containing the eigenvalues $\lambda_i = \lambda_i(\Phi_S^T \Phi_S - \mathrm{Id}) \in \mathbb{R}$, $i = 1, \ldots, n$, of $\Phi_S^T \Phi_S - \mathrm{Id}$. Therefore, it is

$$\sup_{\|x\|_2=1} |\langle (\Phi_S^T \Phi_S - \mathrm{Id})x, x \rangle| = \sup_{\|x\|_2=1} |\langle DUx, Ux \rangle|$$

$$= \sup_{\|x\|_2=1} |\langle Dx, x \rangle|$$

$$= \sup_{\|x\|_2=1} \left| \sum_{i=1}^{n} \lambda_i |x_i|^2 \right|$$

$$= \max_{i=\{1,\ldots,n\}} |\lambda_i|.$$

The last equality follows by letting $x = e_{i_0}$ be the unit vector corresponding to the index i_0 where $|\lambda_i|$ is maximal, which leads

$$\left| \sum_{i=1}^{n} \lambda_i |x_i|^2 \right| = |\lambda_{i_0}| = \max_{i=\{1,\ldots,n\}} |\lambda_i|.$$

Since (4.55) holds for all $S \subset \{1, \ldots, n\}, |S| \leqslant s$ and since the RIP constant δ_s is the smallest δ such that (4.55) is fulfilled, it is

$$
\begin{aligned}
\delta_s &= \max_{S \subset \{1,\ldots,n\}, |S| \leqslant s} \max_{i \in \{1,\ldots,n\}} |\lambda_i(\Phi_S^T \Phi_S) - 1| \\
&= \max_{S \subset \{1,\ldots,n\}, |S| \leqslant s} \max_{i \in \{1,\ldots,n\}} |\lambda_i(\Phi_S^T \Phi_S - \mathrm{Id})| \leqslant \delta.
\end{aligned}
\tag{4.56}
$$

The equality in (4.56) means that the restricted isometry constant δ_s is that eigenvalue of all possible $\binom{n}{s}$ multiplications $\Phi_S^T \Phi_S$ or singular value of Φ_S where $|S| \leqslant s$ with maximum distance to 1. Now, as it has been intensively discussed in [26] and [83], the maximum and minimum eigenvalues of Gram matrices $X^T X$, where the matrix X is a Gaussian random matrix, tend to have an asymmetric deviation from 1. This finding suggests the following definition of an *asymmetric restricted isometry property*:

Definition 4.6.1 (Asymmetric restricted isometry property [26, Definition 2.1]). For a matrix $\Phi \in \mathbb{R}^{m \times n}$ the *asymmetric restricted isometry property* (A-RIP) constants $\mathcal{L} = \tilde{\mathcal{L}}(\Phi)$ and $\mathcal{U} = \tilde{\mathcal{U}}(\Phi)$ are defined as follows:

$$
\begin{aligned}
\tilde{\mathcal{L}} &:= \min_{\delta \geqslant 0} \delta \ \text{ subject to } \ (1 - \delta)\|x\|_2^2 \leqslant \|\Phi x\|_2^2, \text{ for all } x \in \Sigma_s, \\
\tilde{\mathcal{U}} &:= \min_{\delta \geqslant 0} \delta \ \text{ subject to } \ (1 + \delta)\|x\|_2^2 \geqslant \|\Phi x\|_2^2, \text{ for all } x \in \Sigma_s.
\end{aligned}
\tag{4.57}
$$

By definition, the (symmetric) RIP constant δ_s is the maximum of both asymmetric RIP constants, $\delta_s = \max\{\tilde{\mathcal{L}}, \tilde{\mathcal{U}}\}$. By taking the square root in (4.57) and setting $\mathcal{L} := (1 - \tilde{\mathcal{L}})^{1/2}$ as well as $\mathcal{U} := (1 + \tilde{\mathcal{U}})^{1/2}$, (4.57) can be rewritten as

$$
\mathcal{L}(\Phi)\|x\|_2 \leqslant \|\Phi x\|_2 \leqslant \mathcal{U}(\Phi)\|x\|_2.
\tag{4.58}
$$

With regard to the underlying work it is referred to the special case when the matrix Φ is invertible. Then, one can calculate its condition number by $\kappa(\Phi) = \max_{\|x\|_2=1} \|\Phi x\|_2 / \min_{\|x\|_2=1} \|\Phi x\|_2$. In light of (4.58) this motivates to bound the fraction $\mathcal{U}/\mathcal{L}$ as

$$
\xi(\Phi) := \frac{\mathcal{U}(\Phi)}{\mathcal{L}(\Phi)} \leqslant \frac{\max_{\|x\|_2=1} \|\Phi x\|_2}{\min_{\|x\|_2=1} \|\Phi x\|_2} = \kappa(\Phi),
\tag{4.59}
$$

where the value $\xi(\Phi)$ is called the *restricted condition number* of Φ. Even though there has been done analysis with the form of Definition 4.6.1 [26, 27], in this thesis only the just mentioned special case will be incorporated in which the restricted condition number can be bounded by the usual condition number.

5 | Imaging mass spectrometry in a nutshell

This chapter will give a short introduction to the field of imaging mass spectrometry as it is the basis for the further investigations under both the *data compression* aspect (Chapter 6) and the *compressed sensing* aspect (Chapter 7). Among the many mass spectrometry techniques, the focus lies on *matrix-assisted laser desorption/ionization* (MALDI) as the ion source. In addition, identification of the triggered particles is based on their *time-of-flight* (TOF), meaning that a correlation between their masses and their TOF is used. Both the ion source and the particles determination together are called MALDI-TOF and within this work shortly named MALDI. The presentation of the content in this chapter is based on the articles [7, 181, 192].

5.1. Mass spectrometry

Suppose it is the task to determine the chemical composition of a chemical or biological sample or tissue. A way to accomplish this is to make use of the widespread technique called *mass spectrometry* (MS), from which it is known for having "the capacity to generate more structural information per unit quantity than can be determined by any other analytical technique" [205]. Two prominent examples of its application are cancer research [173] and food safety [210]. Another more specific application is to examine the molecular networking of two spatially separated living microbial colonies [203]. For more areas of application, refer to the review article [7]. Generally, in MS one measures the mass-to-charge (m/z) ratio of gas-phase charged molecules (ions) produced by an ionization of

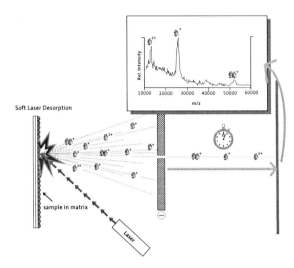

Figure 5.1.: Visualization of the MALDI-TOF principle, taken from [139] (Illustration: Typoform Copyright: The Royal Swedish Academy of Sciences). A tissue sample which is embedded in a matrix is triggered by short laser impulses on several spots. The so dissolved charged molecules are then shortly accelerated followed by a free-flight phase. Depending on how heavy or light the ions are they will reach the detector either sooner or later. Thus, a conclusion to their mass (more precisely, to their mass-to-charge ratio m/z) is possible leading to a mass spectrum for each spot shot.

the molecules from the underlying analyte or component. Among the various mass spectrometry techniques such as the *secondary ion mass spectrometry* (SIMS) [13] and *desorption electrospray ionization* (DESI) [207], this work will only focus on MALDI [113].

The main steps of the MALDI-MS acquisition are visualized in Figure 5.1 and described in the following. MALDI has the advantage that it enables the analysis of heavy molecules. This is achieved through the embedding of the probe in an acid which is generally called *matrix*. The matrix absorbs most of the energy from the laser that acts as an ion source in this context. The ionization of molecules of the sample is triggered by intense short-duration laser pulses on called *spots*. The dissolved ions are then accelerated by an electrostatic field and send to a flight tube. There they are in a free flight phase on their way to a detector and the ions will reach it earlier or later depending on their mass-to-charge ratio (m/z). Within this context one usually talks about the *mass* instead of the mass-to-charge ratio. In addition, the description of the mass is usually followed by the unit specification in *Dalton* [Da] which is another symbol for the unified atomic mass unit[1]. The detector counts the number of ions detected in correlation with the time needed. Based on their TOF, it is then possible to convert to the ions corresponding mass-to-charge ratios or mass values, respectively. As a result for the measurement of one spot one get a so-called *mass spectrum* which shows the chemical composition for this spot within a mass range. There, a high peak in the spectrum indicates a high proportion - usually written in terms of *relative intensity* - of an ion with the associated mass. A typical MALDI dataset is very large and contains 5000-50,000 spectra across 10,000-100,000 m/z-values [7].

The next section deals with the visualization and interpretation of the measured MS data.

5.2. Imaging mass spectrometry

As described above, a mass spectrum is measured for each spot on a grid on the sample. For direct tissue analysis the spots can be transferred to a $\{1, 2, \ldots, n_x\} \times \{1, 2, \ldots, n_y\}$ grid of *pixels* on the tissue, as visualized in Figure 5.2(a) and (b). This directly forms a *hyperspectral datacube*

$$X \in \mathbb{R}_+^{n_x \times n_y \times c},$$

where c corresponds to the number of all given m/z-values. Now, by fixing a m/z-value one can get an individual molecular image or m/z-*image* that represents the spatial distribution of the specific m/z-value, cf. Figure 5.2(c) and (d). This way of visualization, interpretation and also its subsequent analysis of the hyperspectral data X is named *imaging mass spectrometry* (IMS) or *mass spectrometry imaging* (MSI), and in connection with the MALDI system shortly MALDI-IMS [52]. According

---

[1] $1\,\mathrm{u} = 1.660\,538\,921 \times 10^{-27}\mathrm{kg}$. Moreover, in this thesis the declaration with Dalton will be sometimes replaced with a preceding m/z before a mass value.

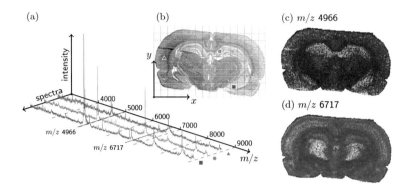

Figure 5.2.: Imaging mass spectrometry data acquired from a rat brain tissue section, adapted from [11]. Each spot on the x, y-grid on the sample in (b) corresponds to one spectrum (a). An m/z-image corresponding to a m/z-value represents the spatial distribution of the ions with this m/z-value, (c) and (d).

to Caprioli [51], "MALDI-IMS enjoys the widest applicability to biological and medical research due to the balance among critical parameters in image analysis, including spatial resolution, molecular types amenable to analysis, molecular mass range, and sensitivity".

The previous observations were related to one single two-dimensional section only. In fact it is possible to consider also three-dimensional data such as a full organ. By cutting the biological sample into n_z consecutive sections and measuring each by the same above mentioned protocol, one gets a datacube $X \in \mathbb{R}_+^{n_x \times n_y \times n_z \times c}$ [14, 174].

5.3. Datasets used in this thesis

5.3.1. Rat brain

The rat brain data was first used for m/z-image segmentation analysis in [9] and amongst others used in [8].

Spectra pre-processing was done in the ClinProTools 2.2 software (Bruker Daltonik): The spectra were baseline-corrected with the mathematical morphology TopHat algorithm (minimal baseline width set to 10%)

[9, 11]. No binning was performed. The spectra were saved into text files and loaded in Matlab R2012a (The Mathworks Inc., Natick, MA, USA). The final dataset comprises 20,185 spectra acquired within the section area of 120×201 pixels, each spectrum of 3045 data bins in the m/z range 2.5 kDa to 10 kDa.

5.3.2. Rat kidney

The rat kidney dataset has first been published by Trede *et al.* in [192] and the following explanation of its properties follows that in [8]. Sample preparation and data acquisition were both done in the MALDI Imaging Lab at the University of Bremen.

Spectra pre-processing was done in the SCiLS Lab software (SCiLS GmbH, Bremen, Germany): The spectra were first baseline-corrected by subtracting a smooth lower envelope curve which consist of wide Gaussians. No binning was performed. The final dataset comprises 6304 spectra acquired within the section area of 113×71 pixels, each spectrum of 10,000 data bins in the m/z range 2 kDa to 18 kDa.

6 | Compression in imaging mass spectrometry

As a consequence of the last chapter about IMS, compression of its data is indispensable due to its typical large size. Therefore, in this chapter several compression techniques according to spatial and spectral redundancies (cf. Section 3.1) in IMS data will be discussed. Among these techniques a new method for peak picking, which has been published in [8], is presented that is more sensitive than other state-of-the art approaches as it selects also peaks with small intensities but with still meaningful information. Moreover, it is shown in a numerical experiment that nonnegative matrix factorization has not only the potential to reduce the IMS data size, but also to extract its most relevant features.

6.1. Introduction

IMS data, as presented in Chapter 5, is a hyperspectral datacube

$$X \in \mathbb{R}_+^{n_x \times n_y \times c}, \tag{6.1}$$

with typical values ranging from $n_x \cdot n_y \in [5000; 50{,}000]$ pixels or m/z-spectra, and about $c \in [10{,}000; 100{,}000]$ channels or m/z-values, respectively. Such large data leads to the following three main problems

1. How can all the data be *stored* and *processed* efficiently?

2. How can the most relevant information be *extracted*?

3. How can the data be *interpreted*?

These problems are still open, and over the last ten to fifteen years research concerning these questions in IMS has rapidly grown [6–11, 14, 18, 19, 51, 55]. As it is stated by Watrous *et al.* [202], "there is still a need to establish standard procedures for data processing [...] for better interpretation of collected data." Moreover, as it has numerically been analyzed in [142] by McDonnell *et al.*, extracting redundancies in IMS data potentially improves its subsequent analysis.

The key for the development of such methods lies, as it is shown in the next sections, in the fact that IMS data is redundant in both the spectral (Section 6.2) as well as spatial (Section 6.3) point of view.

6.2. Peak picking

Mass spectrometry spectra do typically have only very few high intensity peaks compared to the actual dimension of the spectra, cf. Figure 5.2. In other words, m/z-spectra are sparse or compressible alongside the spectral component of the hyperspectral datacube X (6.1). This observation is shown in Figure 6.1 for the rat brain data. There, the first 300 selected peaks with highest intensity are marked in the so-called *mean spectrum*. It is the sum of all single pixel spectra and often used for visualizing the main (peak) characteristics in the data. Due to the compressibility of the spectra it is reasonable to study methods that are able to extract only those relevant peaks or features to compress the data. This leads to the topic of *peak picking* or *feature extraction* [62, 91].

6.2.1. Spectral peak picking

A mathematical way to do peak picking is by deconvolution of the peak-shape of the signal. One efficient way to implement this is given by the *orthogonal matching pursuit* (OMP) algorithm. It was first introduced in [137, 155] in the signal processing context as an improvement of the matching pursuit algorithm [138]. Given an operator A and data g, OMP solves the problem

$$Af = g \tag{6.2}$$

under the the prior knowledge that f is rather sparse in a more general (known) dictionary Ψ than in an orthonormal basis. OMP then iteratively searches for the dictionary elements (atoms) ψ_i that most highly correlate with the residual. This also explains why this method belongs to the class of *greedy* methods, as it "makes a locally optimal choice [ψ_i] in the hope that this choice will lead to a globally optimal solution" [64].

The application of OMP to IMS data has first been studied in [68, 191], where the operator A in (6.2) is given as a deconvolution operator using the Gaussian kernel

$$\kappa(x) = \frac{1}{\pi^{1/4}\sigma^{1/2}} \exp\left(-\frac{x^2}{2\sigma^2}\right), \qquad (6.3)$$

with standard deviation σ. This is motivated from the observation that a given single mass spectrum $g \in \mathbb{R}^c$ can be modeled by the convolution of the mass spectrometry source [68]

$$f = \sum_{i=1}^{c} \alpha_i \delta(\cdot - i) \in \mathbb{R}^c$$

consisting of several Dirac peaks $\delta(\cdot - i)$, $i = 1,\dots,c$, and coefficients $\alpha_i \in \mathbb{R}$, with the Gaussian kernel κ (6.3), i.e.

$$Af := \kappa * f = g.$$

In light of the general formulation (6.2) the problem is then to extract the non-convoluted version f of the mass spectrum g.

For the OMP to use it is natural to use a dictionary Ψ whose columns consist of the shifted Gaussians $\kappa(x - i)$, $i = 0,\dots,c-1$. The algorithm then searches for those elements of Ψ which correlates the most with the peak shapes and intensities in the spectrum g. The algorithm stops if it reaches one of two conditions; either a preset number of peaks are detected, or the norm of the residuum becomes zero or falls below the underlying noise level.

Remark 6.2.1. The just mentioned single spectrum case can also be extended in such a way to search for correlating peaks to all $n_x \cdot n_y$ mass spectra in a datacube X. More precisely, each spectrum would be taken into account by searching for a specified number of high intensity peaks with respect to the full data X and not only to the single spectrum itself. The detection of peaks stops and continues with the next spectrum if

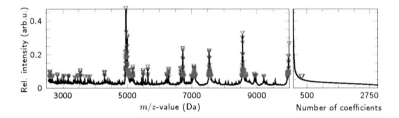

Figure 6.1.: Compressibility of m/z-spectra. The mean spectrum on the left hand side is from the rat brain data (cf. Section 5.3.1). Highlighted with triangles are the first 300 peaks with highest intensity. Following the description of compressibility (cf. (4.1) on page 37), the right hand side clearly shows the quickly decay of the largest coefficients.

a pre-specified number of these is selected. However, since this would increase the number of data points considerably, it is common to only take into account every tenth spectrum in the data [9, 11]. The resulting peak picking then "assigns to each m/z-value a number of spectra in which this m/z-value was selected as a peak." [11]. Furthermore, the resulting peak list can be improved by taking only a certain percentage of all detected peaks. In that way, one can get rid of potentially wrong peaks [11].

Beside the easy applicability of the OMP algorithm, however, it has the weakness that the shape of the Gaussian peaks must be known. Moreover, since the intensity plays a role in the OMP detection of peaks, it probably happens that those with low intensity remain undetected.

6.2.2. Spatial peak picking

Another possible way for detecting the most relevant peaks is to study the m/z-images, i.e. changing from the spectral to the spatial point of view. This is based on the following observation: high intensity peaks in a single spectrum state that there are high intensity pixels at the corresponding m/z-values. Then, similarly to before, high intensity peaks refer to locations of m/z-images with a significant number of high intensity pixels. These images usually exhibit some spatially structured intensity patterns as opposed to those with inherent noise only.

Typically, this selection of structured images is done manually be going through the whole dataset. To work around the less sensitivity of the OMP algorithm and the just mentioned manual selection of m/z-images, a different and automatic approach called *spatial peak picking* has been studied in [8] and will be described next. The idea is to define a measure of structure, or conversely a *measure of chaos*, and to rank the m/z-images according to this value.

As it is mentioned in [8], in literature structure detection techniques typically deal with "finding structured patterns either of a *given* shape [16] or based on edge detection [144], but not with estimating the level of structure." Moreover, to the authors best knowledge only Chubb *et al.* are considering a similar structure detection problem. There, a statistical method is presented for testing the null hypothesis if an image does not contain structure.

The steps for calculating the measure of chaos $\mathcal{C}$ for an m/z-image are visualized in Figure 6.2 and described in the following:

1. Denoising of the i-th m/z-image $X_i := X_{(\cdot,\cdot,i)} \in \mathbb{R}_+^{n_x \times n_y}$ via the bilateral edge-preserving filter $BF(\cdot)$ [190], leading to

$$\tilde{X}_i = BF(X_i).$$

   Bilateral filtering is one method to smoothen an image while preserving its edges. It is motivated from usual Gaussian convolution and "takes into account the difference in value with the neighbors to preserve edges while smoothing" [154].

2. Calculation of the gradient $\nabla \tilde{X}_i \in \mathbb{R}^{n_x \times n_y \times 2}$ and with that the edge detector image $D \in \mathbb{R}^{n_x \times n_y}$,

$$D = \left((\partial_x \tilde{X}_i)^2 + (\partial_y \tilde{X}_i)^2 \right)^{1/2}.$$

   High intensity pixels correspond to sharp changes (discontinuities) of pixel intensities of the original image X_i.

3. Create binary edge mask

$$M \in \{0,1\}^{n_x \times n_y}$$

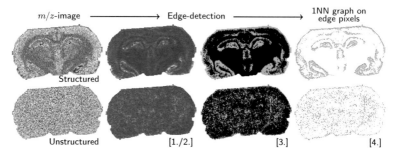

Figure 6.2.: Workflow for the spatial peak picking approach, adapted from [8]. Top and bottom row show the procedure applied on a structured and an unstructured image, respectively. First, denoising on an m/z-image is applied followed by an edge detection step (steps 1.–3.). The detected edge pixels are connected via a 1NN graph (step 4.) whose mean length $\bar{E}$ is calculated. The measure of chaos $\mathcal{C}$ follows as the minimum mean length with respect to all considered quantile values in $\mathcal{Q}$.

by using quantile thresholding on D with parameter $0 < q < 1$. The one-valued pixels in M represent sharp changes of the pixel intensities in X_i.

4. Build a weighted one-nearest neighbor (1NN) graph on the one-valued pixels in M, i.e. link each of those pixels with its closest neighboring one-valued pixel. This part has complexity $\mathcal{O}(p)$ where $p = |\operatorname{supp}(M)|$ [81]. The weight for each graph edge is calculated by the Euclidean distance between the connected pixels.

5. Calculate the mean length $\bar{E} \in \mathbb{R}_+$ of the graph edges, which depends on the set quantile value q, i.e. $\bar{E} = \bar{E}(M(q)) = \bar{E}(q)$.

6. The measure of chaos $\mathcal{C} \in \mathbb{R}_+$ finally results as the minimum mean edge length with respect to several considered quantile values from the set $\mathcal{Q} = \{0.6, 0.65, 0.7, 0.75, 0.8, 0.85, 0.9, 0.95\}$, i.e.

$$\mathcal{C} = \min_{q \in \mathcal{Q}} \bar{E}(q).$$

Remark 6.2.2. The key point in the spatial peak peaking procedure is the calculation of the 1NN graph on the binary edge mask M from step 3. Two of such masks are visualized in the third column of the m/z-images part in Figure 6.2. The upper image has structured intensity patterns. However, the lower one looks rather chaotic.

The idea now is that the connection via a 1NN graph leads to a curve whose mean length is representative for the inherent structure in the image. Such 1NN graphs are visualized in the fourth column of the m/z-images part in Figure 6.2. Suppose, for instance, an image having several single high intensity pixels which are remote from each other. The connection via 1NN then possibly leads to long edges that have an impact on the mean length $\bar{E}$ of all the edges. In contrast, a structured image is supposed to have more shorter than longer edges leading to a lower value $\bar{E}$.

For the statistical evaluation of this method, a set of 250 test m/z-images out of the full rat brain dataset (cf. Section 5.3.1) was manually selected. It consisted of 50 unstructured images and four classes of 50 images, each describing a different type of structure that appears in the data. The classes are divided in images with

i) compact curve-like sets high-intensity pixels ('curves'),

ii) large separated regions of high-intensity pixels ('regions'),

iii) large sets of high-intensity pixels with visible gradients of intensities around them ('gradients'), and

iv) small compact groups of high-intensity pixels ('islets').

Then, for each class the double (or 2-fold) cross-validation accuracy has been calculated when classifying the taken class versus the class of unstructured images. As a classifier, Support Vector Machines have been used [10]. The reader is referred to Kohavi's review article [117] on accuracy estimation methods and the references therein for an introduction to cross-validation.

As a result of this statistical evaluation, it turned out that one can automatically assess between structured and unstructured m/z-images based on the measure of chaos $\mathcal{C}$. As it is evaluated in [8], the best discrimination (accuracy 86%) was achieved for the structured class 'curves' containing m/z-images with compact curve-like series of high-intensity pixels. For the classes of 'curves' and the 'islets' accuracy values of 75% and 71%, respectively, were calculated. The worst discrimination (accuracy 83%) was achieved for the structured class 'gradients' with images with large sets of high-intensity pixels with visible gradients of intensities around them. This can be explained by the fact that the edge detection works well for the 'curves' images and cannot detect clear edges for the 'gradients' images. Using advanced methods for edge detection such as active contours [54] can perhaps help solving this problem.

Remark 6.2.3. It should be noted that there were also corresponding sensitivity and specificity values calculated by Alexandrov *et al.* [8]. They are omitted here to keep the focus on the description of the principle of the spatial structure detection.

An application of the spatial peak picking approach via the measure of chaos $\mathcal{C}$ for the full rat brain dataset is shown in Figure 6.3. The value $\mathcal{C}$ is calculated for each m/z-image following the above steps. Sorting their values in ascending order leads to the graph shown in Figure 6.3(a). Figure 6.3(b) shows the rat brain data mean spectrum and the first 10, 50 and 100 selected m/z-images with lowest measure of chaos. For comparison OMP results with standard deviation $\sigma = 2$ in the Gaussian kernel (6.3) are also included with the 100 strongest correlated peaks. Figures 6.3(c)-(l) present the m/z-images marked in (a). They were selected in equidistant steps based on the measure of chaos. As it is noted in [8], the first seven m/z-images (c)-(i) with low values of the measure of chaos look indeed structured while the last three m/z-images (j)-(l) appear to be unstructured. Moreover, among all 3045 m/z-images, 556 images have lower values of the measure of chaos than that in (j). In Appendix B, similar results are presented for the rat kidney data which is described in Section 5.3.2.

Next to the major peaks in the mean spectrum in Figure 6.3(b) the lower peaks detected with the proposed method are visible. Moreover, the method selected additional peaks compared to OMP such as the large peak at 7100 Da as well as two low intensity peaks at 3625 Da and 4385 Da. The latter peak is highlighted in Figure 6.4 where it is

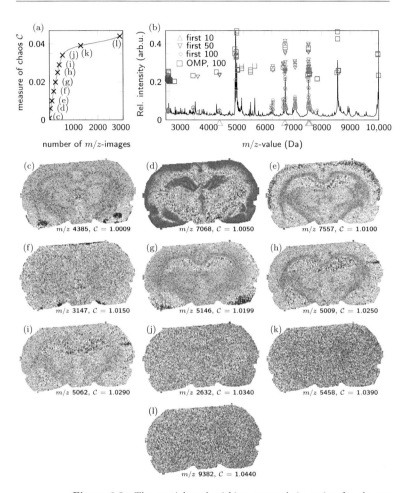

Figure 6.3.: The spatial peak picking approach in action for the rat brain data, adapted from [8]. (a) Sorted values of the measure of spatial chaos for all m/z-images. Crosses indicate ten equidistant values of the measure of chaos; their m/z-images are shown in panels (c)–(l). (b) The dataset mean spectrum with the most 10 (green triangle), 50 (blue triangle), and 100 (orange circle) m/z-images having lowest values of the measure of chaos $\mathcal{C}$. For comparison, 100 peaks selected by the conventional OMP algorithm are shown (red rectangles).

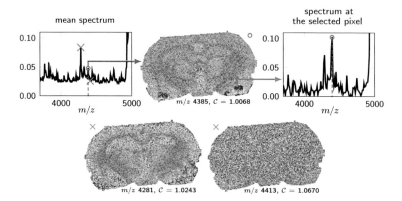

Figure 6.4.: Illustration of the need for sensitive peak picking in IMS. Even though the peak at has low intensity, our method detected it within the first 100 m/z-values by the measure of chaos $\mathcal{C}$. The corresponding m/z-image confirms this selection as it inherent structure. Moreover, a pixel spectrum at a high intensity pixel on this image (shown with an orange arrow) shows a significantly higher peak at the same m/z-value. The red circle indicates the m/z-value detected among 100 m/z-values by our method; green crosses indicate m/z-values not detected but shown for illustration [8].

also shown that this low intensity peak in the mean spectrum in fact correlates to a structured m/z-image. The spectrum acquired at a high-intensity pixel (orange arrow) of this image exhibits a peak of a much higher intensity at the same m/z-value. The other shown m/z-images (marked with green crosses) were not detected within the first 100 m/z-values by the measure of chaos $\mathcal{C}$.

Regarding the processing time for the full rat brain dataset, the proposed method took 16 min, i.e. around 0.31 s for one of its m/z-images on a laptop with i5 2.5 GHz CPU and 8 GB to get the results in Figure 6.3. In contrast, the visual selection of the 20-50 most structured images usually lasts more than one hour.

To conclude this section note that both discussed peak picking methods, the spectral and the spatial one, are able to noticeably reduce the amount of data. In contrast to the OMP method, the spatial peak picking approach using a measure of chaos selects more peaks in the data, leading to a lower compression rate of the data. However, this also gives a better or more sensitive representation of the full data since meaningful smaller peaks are included as well.

6.3. Denoising

In this section a prominent method named total variation minimization [163] is shortly presented which covers spatial redundancies in IMS data. As mentioned in Section 3.1, spatial redundancy refers to correlations between neighboring pixels. Denoising methods are a suitable choice to remove physical noise in an image while preserving its main structure.

The edge-preserving (Rudin, Osher and Fatemi) total variation model [163] is a well-known model for denoising an image $\bar{X} \in \mathbb{R}_+^{n_x \times n_y}$. As it has been explained in Section 3.3.2, the assumption is that the image is piecewise constant or that the image gradient $\nabla \bar{X}$ is sparse ($\|\bar{X}\|_{TV} = \|\nabla \bar{X}\|_1$), leading to the minimization problem

$$\underset{\tilde{X} \in \mathbb{R}_+^{n_x \times n_y}}{\operatorname{argmin}} \frac{1}{2}\|\tilde{X} - \bar{X}\|_F^2 + \alpha \|\tilde{X}\|_{TV}. \tag{6.4}$$

A solution to (6.4) is supposed to be a smoothed version of the probably noisy image $\bar{X}$. This is visualized on an m/z-image in Figure 6.5. The regularization parameter $\alpha > 0$ determines the smoothing effect. The larger α, the greater is this effect.

Motivated from the observations that large variance of noise in individual m/z-images is inherent and that they are sparse with respect to their gradient or piecewise constant respectively, Alexandrov *et al.* proposed in [9] to use an *adaptive* total variation regularization [101]. Instead of choosing the regularization parameter $\alpha > 0$ globally in the minimization problem (6.4), a continuous regularization function

$$\alpha : \mathbb{R}^{n_x \times n_y} \to (0, +\infty)$$

is considered. As the authors in [9] have visually shown, this clearly leads to locally finer regularization results. However for this one needs to calculate a solution $\tilde{X}$ to (6.4) several times to find a suitable local parameter α for all $(x, y) \in \{1, 2, \ldots, n_x\} \times \{1, 2, \ldots, n_y\}$ [101].

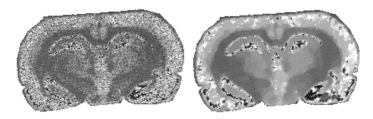

Figure 6.5.: Application of total variation denoising on an m/z-image. Left: Noisy m/z-image from the rat brain dataset. Right: Total variation denoising via (6.4).

In Chapter (7), the optimization problem (6.4) will be solved with respect to a sum of total variation norms over all c m/z-images of the hyperspectral datacube X (6.1). Clearly, in this case finding a global regularization parameter α so that it suits for each of the m/z-images might be difficult, especially when it is known that their inherent noise usually varies from channel to channel [9]. But since the number of channels is usually large, the computational cost of the adaptivity outweigh the benefits. Therefore, only a global parameter choice for α will be used.

6.4. Nonnegative matrix factorization

In the previous two sections the compression in IMS data took place from either the spectral or the spatial perspective. In this section, the idea to apply a matrix factorization of the full data matrix

$$X \in \mathbb{R}_+^{n \times c}, \qquad (6.5)$$

is briefly presented, where X is a reshaped version of the hyperspectral datacube (6.1) with $n = n_x \cdot n_y$. In light of Section 3.3.3, the aim of the factorization can be seen as a way to extract the most frequent m/z-images and their locations which are given in *fingerprint spectra*.

As discussed in Chapter 5, the dimensions of n and c in (6.5) are usually very large. Therefore not only is the interpretation of the data problematic, but also the memory required in a computer to load and work with it can be an issue. Nonnegative matrix factorization (NMF) is a tool which tries to find a much smaller representation of the data X. In that sense it searches for two very small positive matrices $P \in \mathbb{R}_+^{n \times \rho}$ and $S \in \mathbb{R}_+^{\rho \times c}$ where $\rho \ll \min\{n, c\}$ such that

$$X \approx PS. \tag{6.6}$$

As described in Chapter 3, the idea is that the columns of P and the rows of S will in some sense form a basis of the whole data. Put differently, with respect to IMS the ρ rows of S are "characteristic spectra reflecting different metabolites [protein, peptide, etc.]", since the latter are typically "reflected by several isotope patterns" [115, 116]. The spectra are sometimes also simply called *sources* [108], *loadings* (adapted from PCA) [114, 141], *mixtures, fingerprint spectra* or *spectral signatures* [97]. In the context of IMS data the columns of P are called *scores* (adapted from PCA) [114, 141] or *soft segmentation maps* as they divide the measured sample into "regions with different metabolic structures" [116].

The application of matrix factorization to IMS data has been studied in [25, 116] under the often used notion *blind source separation* (BSS). In [179], the general approach is studied using the techniques from [75]. More precisely, the authors in [75] studied the problem geometrically by finding a simplicial cone which contains all data points being in the positive orthant. In [111], NMF results for both 2D and, in particular, 3D datasets were presented. For the numerics, the authors have used the alternating update approach by Lee and Seung [120, 121] as described in Chapter 3. More precisely, one aims in minimizing the functional

$$\frac{1}{2}\|PS - X\|_F^2. \tag{6.7}$$

By incorporating the knowledge on mass spectrometry data, a combination of both the spectral (ℓ_1) and the spatial (TV) approach from the previous two sections might be investigated. With the a priori information that the spectra $S_{(i,\cdot)}$, $i = 1, \ldots, \rho$, are compressible and that the m/z-images $P_{(\cdot,i)}$ are supposed to be piecewise constant; the resulting functional reads

$$\Theta(P, S) = \frac{1}{2}\|PS - X\|_F^2 + \alpha \sum_{i=1}^{\rho} \|P_{(\cdot,i)}\|_{TV} + \beta\|S\|_1, \tag{6.8}$$

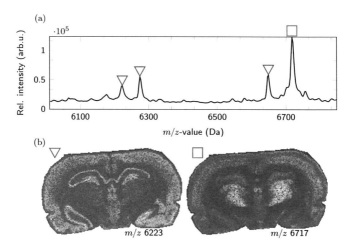

Figure 6.6.: Visualization of the given MALDI test data. (a) presents the mean spectrum. The left image in (b) shows the m/z-image whose structure appears at the m/z locations which are marked with triangles ($\triangledown$) in the mean spectrum in (a). Accordingly, the right image in (b) presents an m/z-image that appears where the square ($\square$) is located.

with the additional positivity constraints $P, S \geqslant 0$.

The expected structure of the solution to (6.8) should mentioned in order to highlight the ability of NMF in applications. It is useful for the reduction of data and localization of metabolites. To investigate these abilities, consider (6.8) with $\rho = 4$ and regularization parameters $\alpha = 500$ and $\beta = 300$ on a part of the rat brain data (cf. Section 5.3.1). It is comprised of $c = 330$ channels covering the m/z-range from 6012 Da to 6870 Da. In light of the form (6.5) one gets the hyperspectral data $X \in \mathbb{R}_+^{24,120 \times 330}$. Figure 6.6 presents the mean spectrum, i.e. the sum over all mass spectra in X. Clearly visible is the fact that the data consists of only two main m/z-images (at 6223 Da and 6717 Da) as well as four highest intensity peaks. It needs to be mentioned that the m/z-images at the triangle positions are of the same structure of the left image in Figure 6.6(b) and differ only slightly in their intensities.

The result of the matrix factorization problem (6.8) is shown in Fig-

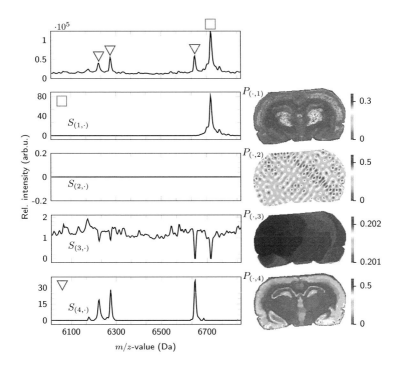

Figure 6.7.: Solution of the minimization problem (6.8) for the described sample rat brain data. First row presents the mean spectrum from the reconstruction $X \approx PS$. The subsequent rows two to five present the respective row of S or the column of P, respectively.

ure 6.7 where it is first recognizable that both the spectral signatures $S_{(i,\cdot)}$ and the soft segmentation m/z-images $P_{(\cdot,i)}$ are positive and that the mean spectrum, given in the first row, is perfectly recovered (cf. Figure 6.6). Furthermore, the second row presents the first signature spectrum $S_{(1,\cdot)}$ as well as one of the main m/z-images in the data given in the first column P. What is notable here is the fact that the m/z-image appears exactly where the peak in $S_{(1,\cdot)}$ arises and nowhere else. The same is visible for the last row in which the pseudo spectrum $S_{(4,\cdot)}$

extracts the locations and the intensities of the corresponding m/z-image in $P_{(\cdot,4)}$. As there are no more structured intensity images left to extract in the matrix factorization, $P_{(\cdot,2)}$ and $P_{(\cdot,3)}$ in the second and the third row are images of different kind. From experience one knows that there exists a visible gradient in the rat brain data due to the MALDI sample preparation. Surprisingly, this is nicely visualized in the third row in $P_{(\cdot,3)}$. The second row contains only a noisy image $P_{(\cdot,2)}$ and as this appears nowhere in the sample data, the coefficients in $S_{(2,\cdot)}$ are accordingly set to zero.

The presented example illustrates the potential of matrix factorization not only to reduce the data tremendously, but also to enable the localization as well as the extraction of metabolites. A detailed examination of nonnegative matrix factorization techniques for IMS data is beyond the scope of this work, but is definitely worthy of further research.

Finally in this chapter it should be mentioned that recently Palmer *et al.* [153] have shown a method to compress the data X in yet another way. It incorporates the very prominent idea to apply *random projections*, which means that the data dimensionality is reduced "by projecting it onto a set of randomly chosen vectors" [153]. In their paper, this is realized on the transposed data $X^T \in \mathbb{R}_+^{c \times n}$ by multiplication with a random matrix $\Omega \in \mathbb{R}^{n \times \rho}$ from the right hand side, i.e.

$$S := X^T \Omega \in \mathbb{R}^{c \times \rho},$$

where $\rho \gtrsim \mathrm{rank}(X^T)$. If Ω is filled with values at random from a certain distribution, the idea is then to create a data basis $Q \in \mathbb{R}^{c \times \rho}$ (more precisely an orthonormal matrix) via QR decomposition from the much smaller size matrix S. Despite the inherent randomness, S is supposed to still inherit all information of X. The compression or projection of the large data X then reads as the following matrix factorization

$$\hat{X} := Q^T X^T \in \mathbb{R}^{\rho \times n},$$

where the resulting compressed data matrix $\hat{X}$ is much smaller than X. In that way, Palmer *et al.* have shown an example where they have reduced original data X with a compression ratio of 0.0056. In other words, the compression led to data that has less than 1% of the original size.

At last note that the full dataset X and therefore each of its m/z-images or pixel spectra can be recreated from Q and $\hat{X}$ via $X^T = Q\hat{X}$ since Q is orthonormal.

6.5. Conclusion

In this chapter, compression techniques which were discussed in Chapter 3 have been applied to IMS data. Indispensable for this was the analysis of the sparsity term in IMS with respect to both the spectral and the spatial dimension. In addition to the known orthogonal matching pursuit (OMP) algorithm for peak picking, a new and more sensitive approach has been presented which is based on a measure for the structuredness of an m/z-image. This method and its numerical analysis was published in [8].

As only a simple gradient for calculating the image edges was used, images with strong noise or those with quite smooth edges could probably be detected as structured, but with a disproportionately high measure. Therefore, future work might investigate other more complex gradient methods such as the active contours approach [54] or even a combination of several together. Furthermore, one might replace the 1NN connecting approach of the pixels in the thresholded image with another type of connecting graph, e.g. with a *minimum spanning tree* (MST).

The very prominent topic of nonnegative matrix factorization (NMF) in IMS has been only shortly investigated. Ongoing research results from [157] illustrated that NMF has great potential to reduce the large amount of IMS data considerably with meaningful feature extraction. Of course, this motivates further research in this direction where one should probably first start with an extensive study of the zoo of different NMF approaches and algorithms.

7 | Compressed sensing in imaging mass spectrometry

In this chapter the main results of this thesis are presented. A model, which is the first of its kind, is introduced that enables compressed sensing (CS) measurements in the field of imaging mass spectrometry (IMS). The model incorporates a priori knowledge on IMS data by making use of both an ℓ_1- and a TV-term. It is shown that recovery of the measured data is robust against noise under certain assumptions on the CS setup. Numerical results finally conclude the chapter. The results of this chapter have been published in [18, 19].

7.1. Introduction

Recall from Chapter 4 that it is, under suitable assumptions on $\Phi \in \mathbb{R}^{m \times n}$ and Ψ, possible to recover the signal x by using the a priori information that the signal $x \in \mathbb{R}^n$ is sparse or compressible in a basis $\Psi \in \mathbb{R}^{n \times n}$. By taking m measurements $y_k = \langle \bar{\varphi}_k, x \rangle$, $k = 1, \ldots, m$, where $\bar{\varphi}_k \in \mathbb{R}^n$ is the k-th row of Φ, it is then possible to recover x from $y \in \mathbb{R}^m$ with the basis pursuit approach, that is, by solving the following convex optimization problem

$$\underset{\lambda \in \mathbb{R}^n}{\operatorname{argmin}} \|\lambda\|_1 \text{ subject to } \|y - \Phi\Psi\lambda\|_2 \leqslant \varepsilon. \tag{7.1}$$

One of the many applications of CS is in hyperspectral imaging. A hardware realization of CS in that hyperspectral situation applying the single-pixel camera [183] has been studied in, for example, [182]. From the theoretical point of view mathematical models have been studied for

CS in hyperspectral image reconstruction under certain priors [97, 98, 100]. Suppose that it is given a hyperspectral datacube $X \in \mathbb{R}^{n_x \times n_y \times c}$ whereas $n_x \times n_y$ denotes the spatial resolution of each image and c the number of channels. As before, by concatenating each image as a vector it is

$$X \in \mathbb{R}^{n \times c} \tag{7.2}$$

with $n := n_x \cdot n_y$. In the context of CS one aims to reduce the number of measurements while still being able to reconstruct the full data (7.2). Often in hyperspectral imaging, this goes hand in hand by taking $m \ll n$ measurements for each spectral channel $1 \leqslant j \leqslant c$ and to formulate a reconstruction strategy based on hyperspectral data priors, see e.g. [98, 100]. In [100], for example, the authors assume the hyperspectral datacube to have low rank and piecewise constant channel images. Therefore the following convex optimization problem is presented

$$\underset{\tilde{X} \in \mathbb{R}^{n \times c}}{\mathrm{argmin}} \|\tilde{X}\|_* + \tau \sum_{j=1}^{c} \|\tilde{X}_j\|_{TV} \text{ subject to } \|Y - \Phi\tilde{X}\|_F \leqslant \varepsilon, \tag{7.3}$$

where $\|\cdot\|_*$ and $\|\cdot\|_{TV}$ denote the nuclear norm (the sum of the singular values) and the TV semi-norm, respectively. Furthermore the notation

$$X_j := \Omega_j X := (\Omega \circ C_j)X, \quad j = 1, \dots, c, \tag{7.4}$$

is used, where C_j maps from a hyperspectral data matrix X to its j-th image in vectorized form and Ω concatenates it to an $n_x \times n_y$ image X_j. This notation will also be used in this chapter. τ is some positive balancing parameter, and the linear operator Φ is the measurement matrix as previously described. The reason for using the nuclear norm as one of the regularization terms arises from the fact that hyperspectral data often has high correlations in both the spatial and the spectral domains.

Another application of CS in hyperspectral imaging is in calculating a compressed matrix factorization or a (blind) source separation of the data $X \in \mathbb{R}^{n \times c}$ (cf. Chapter 6), for example

$$X = SH^T, \tag{7.5}$$

where $S \in \mathbb{R}^{n \times \rho}$ is a so-called *source matrix*, $H \in \mathbb{R}^{c \times \rho}$ is a *mixing matrix* and $\rho \ll \min\{n, c\}$ denotes the number of sources in the data which is supposed to be a priori known. This model has been recently studied in the case of known mixing parameters H of the data X in [99] and with both matrices unknown in [97]. If H is known and if the columns of S are sparse or compressible in a basis Ψ, the problem in [99] becomes

$$\underset{\lambda \in \mathbb{R}^{\bar{n}}}{\operatorname{argmin}} \|\lambda\|_1 \text{ subject to } \|Y - \Phi \bar{H} \Psi \lambda\|_2 \leqslant \varepsilon, \tag{7.6}$$

where $\bar{n} = \rho \cdot n$, $\bar{H} = H \otimes I_n$, with denoting $\otimes$ the usual Kronecker product (cf. (2.2), page 16) and I_n the $n \times n$ identity matrix. The authors in [99] also studied the case where the ℓ_1-norm in (7.6) is replaced by the TV norm with respect to the columns of S, $\sum_{j=1}^{\rho} \|S_j\|_{TV}$, where S_j is defined as in (7.4) with proper dimensions. In this instance, solving (7.6) yields a decomposition as in (7.5), where the columns of S contain the ρ most representative images of the hyperspectral datacube and the rows of H contain the corresponding pseudo spectra.

In this chapter a reconstruction model for hyperspectral data similar to (7.3) and (7.6) is investigated, but with special motivation for IMS data. Let $X \in \mathbb{R}_+^{n \times c}$ be the hyperspectral IMS data and assume that there exists a sparse decomposition of the spectra $X_{(i,\cdot)} \in \mathbb{R}_+^c$ for $i = 1, \ldots, n$ with respect to some basis $\Psi \in \mathbb{R}_+^{c \times c}$, i.e.

$$X^T = \Psi \Lambda \tag{7.7}$$

where $\Lambda \in \mathbb{R}_+^{c \times n}$. In addition assume the m/z-images to be piecewise constant so that their total variation norm is small. As it is explained in more detail in the next Sections 7.3 and 7.4, by applying compressed measurements via $\Phi \in \mathbb{R}^{m \times n}$ and (7.4), the minimization problem then becomes

$$\underset{\Lambda \in \mathbb{R}^{c \times n}}{\operatorname{argmin}} \|\Lambda\|_1 + \sum_{j=1}^{c} \|\Lambda_j\|_{TV}$$
$$\text{subject to } \|Y - \Phi \Lambda^T \Psi^T\|_F \leqslant \varepsilon, \ \Lambda \geqslant 0. \tag{7.8}$$

Following the descriptions from Chapter 6 concerning the sparsity aspects in IMS data, it is known a priori that mass spectra in IMS are typically nearly sparse or compressible. For this reason, the ℓ_1-norm is used as one

101

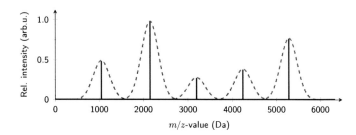

Figure 7.1.: An illustration of a peak-picking approach in mass spectrometry, first published in the proceedings of SampTA 2013 published by EURASIP [19]. Instead of finding a reconstruction $\tilde{X}$ via $\tilde{X}^T = \Psi\tilde{\Lambda}$, it is aimed to directly recover the features $\tilde{\Lambda}$. Dashed line (- - -): Reconstruction of the i-th spectrum, i.e. $\tilde{X}^T_{(i,\cdot)} = (\Psi\tilde{\Lambda})_{(\cdot,i)}$. Solid line (———): Only the main features of the i-th spectrum $\tilde{\Lambda}_{(\cdot,i)}$, i.e. the main peaks, are extracted.

regularization term [8, 68]. The TV-term comes into play because the m/z-images have sparse image gradients [9].

Moreover, with respect to (7.7) and the related work of Louis [130], it is aimed to directly reconstruct the main features Λ from the measured data Y without inverting the operator Ψ with a sparsity constraint as done in [68]. More precisely, one strives for directly reconstructing the features $\tilde{\Lambda}$ from $\tilde{X}^T = \Psi\tilde{\Lambda}$ from only $m \ll n_x \cdot n_y$ measurements, see Figure 7.1. For this reason, in (7.8) the j-th feature m/z-image Λ_j instead of its convoluted version $(\Psi\Lambda)_j$ is TV minimized.

As the full dataset will be reconstructed, there is no need to know a priori the number ρ of the mixing signatures of the underlying data, which makes the here presented approach different from [97, 99].

7.2. The compressed sensing process

Recall from Chapter 5 that IMS data is a hyperspectral datacube consisting of one mass spectrum for each pixel. The length of each spectrum depends on the number $c > 0$ of m/z-channels that have been selected before MS data acquisition. By fixing one specific m/z-value, an m/z-image is given that represents the spatial distribution of the given mass in the biological sample, see Figure 5.2, page 78. More formally, for the dimensions $\{1, \dots, n_x\} \times \{1, \dots, n_y\} \subseteq \mathbb{Z}_+^2$ and $c \in \mathbb{N}_+$, the IMS datacube $X \in \mathbb{R}_+^{n_x \times n_y \times c}$ consists of m/z-images $X_{(\cdot,\cdot;k)} \in \mathbb{R}_+^{n_x \times n_y}$ for $k = 1, \dots, c$ of image resolution $n_x \times n_y$. Since in MALDI measurement process one counts the (relative) number of charged particles of a given mass that reaches the detector, it is natural to assume the data to be non-negative. By reshaping each image as a vector the hyperspectral data becomes as in (7.2)

$$X \in \mathbb{R}_+^{n \times c}$$

where $n := n_x \cdot n_y$, so that each column in X corresponds to one m/z-image and each row corresponds to one spectrum.

As described in Chapter 5, part of the IMS measurement process consists of the ionization of the given sample. In MALDI-IMS, for instance, the tissue is ionized by a laser beam, which hits each of the n pixel of a predefined grid, producing n independently measured spectra. The main goal here is to use the theory of compressed sensing [45–47, 49, 72] to reduce the number of spectra required but still be able to reconstruct a full MALDI-IMS datacube X.

In the context of compressed sensing (cf. Chapter 4), each entry y_{ij} of the measurement vectors $y_i \in \mathbb{R}^c$ for $i = 1, \dots, m$ and $j = 1, \dots, c$ is the result of an inner product between the data $X \in \mathbb{R}_+^{n \times c}$ and a test function $\varphi_i \in \mathbb{R}^n$ with components φ_{ik}, i.e.

$$y_{ij} = \langle \varphi_i, X_{(\cdot,j)} \rangle. \tag{7.9}$$

From the IMS perspective these y_i for $i = 1, \dots, m$ are called the *measurement-mean spectra* since they are calculated by the mean intensities on each channel, see Figure 7.2. This can be seen by rewriting (7.9) as

$$
\begin{aligned}
y_i^T = \varphi_i^T X \\
= (\varphi_{1i} X_{11} + \dots + \varphi_{in} X_{n1}) + (\varphi_{i1} X_{12} + \dots + \varphi_{in} X_{n2}) + \dots \\
+ (\varphi_{i1} X_{1c} + \dots + \varphi_{in} X_{nc}) \\
= \varphi_{i1}(X_{11} + \dots + X_{1c}) + \dots + \varphi_{in}(X_{n1} + \dots + X_{nc}) \\
= \sum_{k=1}^{n} \varphi_{ik} X_{(k,\cdot)},
\end{aligned}
\tag{7.10}
$$

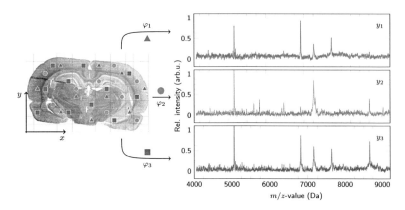

Figure 7.2.: Compressed sensing measurements in imaging mass spectrometry. Each measurement φ_i on the sample (left) (triangle (▲), circle (●), square (■)) leads to a measurement y_i (right) called a measurement-mean spectrum.

which directly shows that the measurement vectors y_i^T are linear combinations of the original spectra $X_{(k,\cdot)}$. In matrix form (7.9) or (7.10) becomes

$$Y = \Phi X \in \mathbb{R}^{m \times c}, \qquad (7.11)$$

where $\Phi \in \mathbb{R}^{m \times n}$ is the measurement matrix. One is looking for a reconstruction of the data X based on these m measurement-mean spectra, each measured by one linear function φ_i. Clearly, by (7.10), each row in Y can be interpreted as a measurement-mean spectrum. By incorporating inherent noise $Z \in \mathbb{R}_+^{m \times c}$ with $\|Z\|_F \leqslant \varepsilon$ that arises during the mass spectrometry measurement process, (7.11) becomes

$$Y = \Phi X + Z \in \mathbb{R}^{m \times c}. \qquad (7.12)$$

Remark 7.2.1. Here it is explicitly assumed to have Gaussian noise for simplicity, but it should be noted that perhaps a Poisson noise framework [9, 158] or a sum of both where the Gaussian part dominates [12] might be more suitable to IMS data. The latter guess is boost with the results known from liquid chromatography-mass spectrometry (LC-MS), where noise analysis yields to similar models [12].

Finding a reconstruction of the data X from the measurements Y in (7.11) is hopeless due the ill-posed nature of the problem. Therefore, additional a priori knowledge is needed to find at least those reasonable solutions which also fulfills the given data properties. To remedy this, the next two subsections recall the in Chapter 6 presented two notions of sparsity that arise in imaging mass spectrometry.

7.3. First assumption: compressible spectra

For each pixel in the sample a mass spectrum with positive real entries is obtained, i.e.

$$X_{(k,\cdot)} \in \mathbb{R}^c_+, \quad k = 1, \ldots, n.$$

As it was motivated in Figure 6.1 on page 84, IMS spectra are compressible in spectral domain. It is therefore assumed that these spectra are sparsely presented by a suitable choice of functions $\psi_i \in \mathbb{R}^c_+$ for $i = 1, \ldots, c$, cf. Figure 7.3. More concretely, this means that there exists a matrix $\Psi \in \mathbb{R}^{c \times c}_+$ such that for each spectrum $X_{(k,\cdot)}$ there is a coefficient vector $\lambda_k \in \mathbb{R}^c_+$ with $\|\lambda_k\|_0 \ll c$, such that

$$X^T_{(k,\cdot)} = \Psi \lambda_k, \quad k = 1, \ldots, n. \tag{7.13}$$

In this thesis it is assumed that the basis functions are shifted Gaussians [9, 63, 68, 70]

$$\psi_k(x) = \frac{1}{\pi^{1/4} \sigma^{1/2}} \exp\left(-\frac{(x-k)^2}{2\sigma^2} \right), \tag{7.14}$$

where the standard deviation σ needs to be set based on the data [9]. However, in matrix form, the sparsity property (7.13) can be written as

$$X^T = \Psi \Lambda, \tag{7.15}$$

where $\Lambda \in \mathbb{R}^{c \times n}_+$ is the coefficient matrix. The single-spectrum case from (7.13) can simply be found in (7.15): One column in X^T corresponds to one spectrum. The multiplication of Ψ with one column of Λ is exactly the same as in (7.13). However, in light of the compressible spectra, the aim should be to minimize each column $\Lambda_{(\cdot, i)}$ of Λ with respect to the

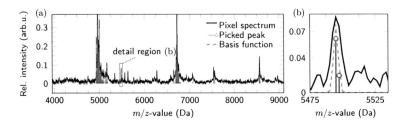

Figure 7.3.: An example of a pixel spectrum from the rat brain sample dataset and its peak-picking result via ℓ_1 minimization with Gaussians as basis elements $\psi_k(\cdot)$ with $\sigma = 0.75$ (see (7.14)), (a). (b) presents a detailed view of the marked region in (a), in which a basis element as well as the coefficients of the detected peak are visualized.

ℓ_0-"norm", since each represents the sparse peak-list information based on Ψ. Thus, for one spectrum one has $\|\Lambda_{(\cdot,i)}\|_0$ and for all spectra this reads

$$\|\Lambda\|_0. \tag{7.16}$$

Note that the notation from spectra and images (the order in the brackets in the index) changes for Λ due to the transposition in (7.15).

Putting (7.12) and (7.15) together leads to

$$Y = \Phi\Lambda^T\Psi^T + Z. \tag{7.17}$$

7.4. Second assumption: sparse image gradients

By fixing one m/z-value $i_0 \in \{1, \ldots, c\}$ it results in a vector $X_{(\cdot,i_0)} \in \mathbb{R}_+^n$ (one column of the dataset X), which by (7.4) is also an m/z-image $X_{i_0} \in \mathbb{R}_+^{n_x \times n_y}$ that represents the spatial distribution of the fixed mass m_0 in the measured biological sample. It is a priori known from Section 6.3 that these m/z-images are sparse with respect to their gradient. In addition, it is also known about the large variance of noise inside each individual m/z-image [9]. This motivates to minimize each m/z-image with respect to its TV semi-norm.

The matrix Ψ is columnwise filled with the shifted Gaussian kernels from (7.14) and it can therefore be interpreted as a convolution operator in the spectral (m/z) dimension. With respect to (7.15), this means that the spectra $X_{(k,\cdot)}$, $k = 1, \ldots, n$, are only sums of the shifted Gaussian kernels, see Figure 7.1. Hence, the multiplication of Ψ in (7.15) does not effect the structure of each m/z-image X_i. Therefore, instead of minimizing

$$\|X_i\|_{TV} = \|(\Psi\Lambda)_i^T\|_{TV},$$

one concludes that it suffices to minimize the TV norm of the c images given through the coefficients Λ, i.e.

$$\sum_{i=1}^{c} \|\Lambda_i\|_{TV}. \tag{7.18}$$

7.5. The final model

In total, it is now possible to formulate a first model for CS in IMS. Wanted is a positiv coefficient matrix $\tilde{\Lambda} \in \mathbb{R}_+^{c \times n}$ such that

1. the reconstructed datacube $\tilde{X}^T = \Psi\tilde{\Lambda}$ is consistent with the observed measurements Y up to a certain noise level ε, see (7.17)

2. the m/z-images X_i for $i = 1, \ldots, c$ or, more precisely, the deconvoluted analogs Λ_i, have sparse image gradients, see (7.18)

3. each spectrum $X_{(i,\cdot)}$ can be represented by only a few peaks indicating sparse coefficient vectors $\Lambda_{(\cdot,i)}$

This leads to the following optimization problem

$$\operatorname*{argmin}_{\Lambda \in \mathbb{R}^{c \times n}} \|\Lambda\|_0 + \sum_{i=1}^{c} \|\Lambda_i\|_{TV}$$
$$\text{subject to } \|Y - \Phi\Lambda^T\Psi^T\|_F \leqslant \varepsilon, \ \Lambda \geqslant 0. \tag{7.19}$$

It turns out that minimizing with respect to the ℓ_0-"norm" is NP-hard [148]. Furthermore, this norm is not convex. To obviate this it is common

107

to replace this norm with the ℓ_1-norm [48, 74]. To simplify the following presentation, the linear mapping

$$\mathcal{D}_{\Phi,\Psi} : \mathbb{R}_+^{c\times n} \to \mathbb{R}^{m\times c}, \Lambda \mapsto \Phi\Lambda^T\Psi^T \qquad (7.20)$$

is introduced, whereby (7.19) becomes

$$\operatorname*{argmin}_{\Lambda\in\mathbb{R}^{c\times n}} \|\Lambda\|_1 + \sum_{i=1}^{c} \|\Lambda_i\|_{TV} \qquad (7.21)$$
$$\text{subject to } \|Y - \mathcal{D}_{\Phi,\Psi}\Lambda\|_F \leqslant \varepsilon, \Lambda \geqslant 0.$$

7.6. Robust recovery

In this section it will be shown that the ℓ_1-reconstruction in spectral and image domain of the unknown matrix $\Lambda \in \mathbb{R}_+^{c\times n}$ in (7.21) is robust with respect to noise. In the case of compressed MALDI hyperspectral imaging this means that the pixel spectra as well as the m/z-images are stably reconstructed. For this it is needed to generalize the results from [47, 150, 151].

The robustness result presented in this section combines both the respective ℓ_1- and the TV-recovery guarantees. It takes into account both the sparsity in the spectral dimension as well as in the spatial dimension. To the best knowledge of the author of this thesis, guarantees have so far only be considered with respect to the sparsity in the spatial dimensions in hyperspectral data [99, 100, 150]. In addition it is, similar to [150, 151], assumed to have measurements on the image gradients.

Remark 7.6.1. Note that the analysis in this section will not take place on a hyperspectral datacube $X \in \mathbb{R}_+^{n_x\times n_y\times c}$ but on its reshaped data matrix version $X \in \mathbb{R}_+^{n\times c}$ with c m/z-images $X_j \in \mathbb{R}^{n_x\times n_y}$.

The reader is reminded of the notation introduced in (7.4),

$$X_j := \Omega_j X := (\Omega \circ C_j)X, \quad j = 1,\dots,c, \qquad (7.22)$$

where C_j maps from a hyperspectral data matrix X to its j-th image in vectorized form and Ω concatenates it to an $n_x \times n_y$ image X_j. Whenever the notation Z_j appears in the sense of an image from a matrix $Z \in \mathbb{R}_+^{n\times c}$ it is meant that, formally, the procedure (7.22) should be in mind.

As it was described in Chapter 4, one of the fundamental ideas in CS is the following restricted isometry property (RIP) and its derivatives such as the *dictionary restricted isometry property* (D-RIP) and the *asymmetric isometry property* (A-RIP). For the sake of completeness and readability the definitions from Chapter 4 shall be extended here for the multichannel IMS context.

Definition 7.6.2. The linear operator $\mathcal{A} : \mathbb{R}^{n_x \times n_y} \to \mathbb{R}^{m \times p}$ has the *restricted isometry property* of order s and level $\delta \in (0, 1)$ if

$$(1 - \delta)\|X\|_F^2 \leqslant \|\mathcal{A}X\|_F^2 \leqslant (1 + \delta)\|X\|_F^2$$

for all s-sparse $X \in \mathbb{R}^{n_x \times n_y}$. The smallest δ for which this holds is the *restricted isometry constant* for the operator $\mathcal{A}$ and is denoted by δ_s.

Definition 7.6.3. A matrix $D \in \mathbb{R}^{n_x \times n_x}$ satisfies the *asymmetric restricted isometry property* of order s, if for all s-sparse $X \in \mathbb{R}^{n_x \times n_y}$ the following inequalities hold:

$$\mathcal{L}(D)\|X\|_F \leqslant \|DX\|_F \leqslant \mathcal{U}(D)\|X\|_F,$$

where $\mathcal{L}(D)$ and $\mathcal{U}(D)$ are the largest and the smallest constants for which the above inequalities hold. The *restricted condition number* of D is defined as

$$\xi(D) = \frac{\mathcal{U}}{\mathcal{L}}.$$

The dictionary RIP (D-RIP) extends the notion of the standard RIP to matrices adapted to a dictionary.

Definition 7.6.4. A linear operator $\mathcal{A} : \mathbb{R}^{d_1 \times d_2} \to \mathbb{R}^{m \times p}$ has the *dictionary restricted isometry property* of order s and level $\delta^* \in (0, 1)$, adapted to a dictionary $D \in \mathbb{R}^{d_1 \times d_1}$, if for all s-sparse $X \in \mathbb{R}^{d_1 \times d_2}$ it holds

$$(1 - \delta^*)\|DX\|_F^2 \leqslant \|\mathcal{A}DX\|_F^2 \leqslant (1 + \delta^*)\|DX\|_F^2.$$

The robustness result that will be shown in Theorem 7.6.8 rests mainly on the following proposition. They are both generalizations of Proposition 4.4.2, page 61 and Theorem 4.4.7, page 66, from [150, 151]. The proposition states, that if a family of noisy D-RIP-measurements fulfills generalized cone and tube constraints as introduced in [151], then robust recovery is possible.

Remark 7.6.5. The following proposition is not restricted to images $X_1, \ldots, X_m$ having all the same size $n_x \times n_y$. On the contrary, it even allows that each of the images have its own size $n_x^i \times n_y^i$ for $i = 1, \ldots, m$.

Proposition 7.6.6. *For* $i = 1, \ldots, m$ *fix the parameters* $\varepsilon_i \geqslant 1$, $\sigma > 0$, $\gamma \geqslant 1$, $\delta_i^* < 1/3$, $C_i > 0$ *as well as* $k_i, n_x, n_y, t, p \in \mathbb{N}$. *Suppose that every operator*

$$\mathcal{A}_i : \mathbb{R}^{n_x^i \times n_y^i} \to \mathbb{R}^{t \times p}$$

satisfies the D-RIP of order $5k_i\gamma^2$ *and level* δ_i^*, *that each given dictionary* $D_i \in \mathbb{R}^{n_x^i \times n_x^i}$ *satisfies the A-RIP of order* $5k_i\gamma^2$ *with constants* $\mathcal{L}(D_i)$ *and* $\mathcal{U}(D_i)$ *and suppose that each image*

$$D_i X_i \in \mathbb{R}^{n_x^i \times n_y^i}$$

satisfies a tube constraint

$$\|\mathcal{A}_i D_i X_i\|_F \leqslant C_i \varepsilon_i. \tag{7.23}$$

Set $C := \max C_i$, $\delta^* = \max \delta_i^*$, $\varepsilon := \max \varepsilon_i$, $k := \min k_i$, $K := \max k_i$, $\mathcal{U} := \max \mathcal{U}(D_i)$, $\mathcal{L} := \min \mathcal{L}(D_i)$, $\xi = \mathcal{U}/\mathcal{L}$, *and suppose that* $\gamma \geqslant \xi\sqrt{K}/\sqrt{k} \geqslant 1$. *Further suppose that for each subset* S_i *of cardinality* $|S_i| \leqslant k_i$ *for* $i = 1, \ldots, m$ *a cone constraint of the form*

$$\sum_{i=1}^{m} \|X_{i_{S_i^C}}\|_1 \leqslant \sum_{i=1}^{m} \|X_{i_{S_i}}\|_1 + \sigma \tag{7.24}$$

is satisfied, where $X_{i_{S_i^C}}$ *and* $X_{i_{S_i}}$ *denotes the matrix* $X_i \in \mathbb{R}^{n_x^i \times n_y^i}$ *restricted to the index set* S_i^C *and* S_i, *respectively. Then*

$$\sum_{i=1}^{m} \|X_i\|_F \lesssim m\varepsilon + \frac{\sigma}{\sqrt{K}} \tag{7.25}$$

and

$$\sum_{i=1}^{m} \|X_i\|_1 \lesssim \sqrt{K}m\varepsilon + \sigma. \tag{7.26}$$

Proof. Let $s_i = k_i \gamma^2$ and let $S_i \subset \{1, \ldots, N_i\}$, with $N_i = n_x^i \cdot n_y^i$, be the support set of an arbitrary s_i-term approximation. For each image X_i, $i = 1, \ldots, m$, decompose its complement $S_i^C = \{1, \ldots, N_i\} \backslash S_i$ as

$$X_{i_{S_i^C}} = X_{i_{S_i^1}} + X_{i_{S_i^2}} + \ldots + X_{i_{S_i^{r_i}}} \qquad \text{where} \qquad r_i = \left\lfloor \frac{N_i}{4 s_i} \right\rfloor.$$

Note that $X_{i_{S_i^1}}$ consists of the $4 s_i$ largest magnitude components, i.e the four largest coefficients in 1-norm of X_i over S_i^C. $X_{i_{S_i^2}}$ then consists of the $4 s_i$ largest magnitude components of X_i over $S_i^C \backslash S_i^1$ and so on. By definition, the average magnitude of the nonzero components of $X_{i_{S_i^{j-1}}}$ is larger than the magnitude of each of the nonzero components of $X_{i_{S_i^j}}$, i.e.

$$|X_{i_{S_i^j}}^{(t)}| \leqslant \frac{\sum_{p=1}^{4 s_i} |X_{i_{S_i^{j-1}}}^{(p)}|}{4 s_i}.$$

Therefore, for $j = 2, 3, \ldots, r_i$ it follows

$$\begin{aligned}
\|X_{i_{S_i^j}}\|_F^2 = \sum_{p=1}^{4 s_i} |X_{i_{S_i^j}}^{(p)}|^2 &\leqslant \sum_{p=1}^{4 s_i} |X_{i_{S_i^j}}^{(p)}| \sum_{p=1}^{4 s_i} |X_{i_{S_i^j}}^{(p)}| \\
&\leqslant \|X_{i_{S_i^{j-1}}}\|_1 \frac{1}{4 s_i} \sum_{p=1}^{4 s_i} |X_{i_{S_i^j}}^{(p)}| \\
&\leqslant \|X_{i_{S_i^{j-1}}}\|_1 \frac{1}{4 s_i} \sum_{p=1}^{4 s_i} |X_{i_{S_i^{j-1}}}^{(p)}| \\
&= \frac{1}{4 s_i} \|X_{i_{S_i^{j-1}}}\|_1^2,
\end{aligned}$$

and thus

$$\|X_{i_{S_i^j}}\|_F \leqslant \frac{\|X_{i_{S_i^{j-1}}}\|_1}{2 \sqrt{s_i}}, \qquad j = 2, 3, \ldots, r_i.$$

111

Together with the cone constraint (7.24), one obtains

$$\sum_{i=1}^{m}\sum_{j=2}^{r_i}\|X_{i_{S_i^j}}\|_F \leqslant \sum_{i=1}^{m}\sum_{j=2}^{r_i}\frac{\|X_{i_{S_i^{j-1}}}\|_1}{2\sqrt{s_i}}$$

$$= \sum_{i=1}^{m}\frac{\|X_{i_{S_i^C}}\|_1}{2\gamma\sqrt{k_i}}$$

$$\leqslant \frac{1}{2\gamma\sqrt{k}}\sum_{i=1}^{m}\|X_{i_{S_i}}\|_1 + \frac{\sigma}{2\gamma\sqrt{k}}$$

$$\leqslant \frac{\sqrt{K}}{2\gamma\sqrt{k}}\sum_{i=1}^{m}\|X_{i_{S_i}}\|_F + \frac{\sigma}{2\gamma\sqrt{k}}.$$

In combination with the tube constraints (7.23), the D-RIP for each $\mathcal{A}_i$ as well as the A-RIP for each dictionary D_i, it follows

$$Cm\varepsilon \geqslant \sum_{i=1}^{m}\|\mathcal{A}_i D_i X_i\|_F$$

$$\geqslant \sum_{i=1}^{m}\|\mathcal{A}_i D_i(X_{i_{S_i}} + X_{i_{S_i^1}})\|_F - \sum_{i=1}^{m}\sum_{j=2}^{r_i}\|\mathcal{A}_i D_i X_{i_{S_i^j}}\|_F$$

$$\geqslant \sum_{i=1}^{m}\sqrt{1-\delta_i^*}\|D_i(X_{i_{S_i}} + X_{i_{S_i^1}})\|_F - \sum_{i=1}^{m}\sum_{j=2}^{r_i}\sqrt{1+\delta_i^*}\|D_i X_{i_{S_i^j}}\|_F$$

$$\geqslant \mathcal{L}\sqrt{1-\delta^*}\sum_{i=1}^{m}\|X_{i_{S_i}} + X_{i_{S_i^1}}\|_F$$

$$- \mathcal{U}\sqrt{1+\delta^*}\left(\frac{\sqrt{K}}{2\gamma\sqrt{k}}\sum_{i=1}^{m}\|X_{i_{S_i}}\|_F + \frac{\sigma}{2\gamma\sqrt{k}}\right)$$

$$\geqslant \left(\mathcal{L}\sqrt{1-\delta^*} - \mathcal{U}\frac{\sqrt{K}}{2\gamma\sqrt{k}}\sqrt{1+\delta^*}\right)\sum_{i=1}^{m}\|X_{i_{S_i}} + X_{i_{S_i^1}}\|_F$$

$$- \mathcal{U}\sqrt{1+\delta^*}\frac{\sigma}{2\gamma\sqrt{k}}.$$

Further calculations require that the bracket term is strictly positive, or

$$\sqrt{1-\delta^*} - \xi\frac{\sqrt{K}}{2\gamma\sqrt{k}}\sqrt{1+\delta^*} > 0.$$

With $\gamma \geqslant \xi\sqrt{K}/\sqrt{k}$ it is sufficient to have $\delta^* < 1/3$. For this it follows[1]

$$\sum_{i=1}^{m} \|X_{i_{S_i}} + X_{i_{S_i^1}}\|_F \leqslant 5\frac{C}{\mathcal{L}}m\varepsilon + 3\xi\frac{\sigma}{\gamma\sqrt{k}}.$$

Because of the inequality

$$\sum_{i=1}^{m} \left\|\sum_{j=2}^{r_i} X_{i_{S_i^j}}\right\|_F \leqslant \sum_{i=1}^{m}\sum_{j=2}^{r_i} \|X_{i_{S_i^j}}\|_F$$

$$\leqslant \frac{\sqrt{K}}{2\gamma\sqrt{k}}\sum_{i=1}^{m}\|X_{i_{S_i}} + X_{i_{S_i^1}}\|_F + \frac{\sigma}{2\gamma\sqrt{k}}$$

$$\leqslant \frac{1}{2\xi}\sum_{i=1}^{m}\|X_{i_{S_i}} + X_{i_{S_i^1}}\|_F + \frac{\sigma}{2\gamma\sqrt{k}}$$

and the fact that $\xi \geqslant 1$, it follows that

$$\sum_{i=1}^{m}\|X_i\|_F \leqslant 8\frac{C}{\mathcal{L}}m\varepsilon + 5\xi\frac{\sigma}{\gamma\sqrt{k}} \leqslant 8\frac{C}{\mathcal{L}}m\varepsilon + 5\frac{\sigma}{\sqrt{K}}$$

which proves (7.25). Using the cone constraint (7.24), one concludes that

$$\sum_{i=1}^{m}\|X_i\|_1 = \sum_{i=1}^{m}\|X_{i_S} + X_{i_{S^C}}\|_1$$

$$\leqslant \sum_{i=1}^{m}\|X_{i_{S_i}}\|_1 + \sum_{i=1}^{m}\|X_{i_{S_i^C}}\|_1$$

$$\leqslant 2\sum_{i=1}^{m}\|X_{i_{S_i}}\|_1 + \sigma$$

$$\leqslant 2\sqrt{K}\left(8\frac{C}{\mathcal{L}}m\varepsilon + 5\frac{\sigma}{\sqrt{K}}\right) + \sigma,$$

which proves (7.26). □

---

[1]With $\gamma \geqslant \xi\sqrt{K}/\sqrt{k}$ and $\delta^* < 1/3$ one computes $\sqrt{1-\delta^*} - \xi\frac{\sqrt{K}}{2\gamma\sqrt{k}}\sqrt{1+\delta^*} > \sqrt{1-\delta^*} - \sqrt{1+\delta^*}/2 > 1/5$.

With this proposition in hand it is now possible to formulate the robust recovery result. But before that the reader should be reminded by the following remark on the used notation described in Section 4.4.

Remark 7.6.7. The notations Φ_0 and Φ^0 for a matrix $\Phi \in \mathbb{R}^{(n_x-1)\times n_y}$ (cf. Definition 4.41, page 64) describe augmented matrices by concatenating a row of zeros at the bottom or on top of Φ, respectively.

Based on this definition, it could be shown in Lemma 4.4.3, page 64, that horizontal and vertical gradient measurements can be modeled as a sum of two scalar product expressions using such padded matrices.

As it was then motivated in the Remarks 4.4.5 and 4.4.6 on page 65, this observation led to the two operators

$$\mathcal{A} : \mathbb{R}^{(n-1)\times n} \to \mathbb{R}^{m_1} \quad \text{and} \quad \mathcal{A}' : \mathbb{R}^{(n-1)\times n} \to \mathbb{R}^{m_1},$$

each describing m measurements of a matrix $Z \in \mathbb{R}^{(n-1)\times n}$.

Theorem 7.6.8 (Robust recovery). *For $i = 1,\dots,c$ consider n_x, n_y, c, k_0, k_i, m_1, $m_2 \in \mathbb{N}$, $\gamma \geqslant 1$, $K := \max\{\sum_{i=1}^{c} k_i, k_0\}$ as well as $k := \min\{\sum_{i=1}^{c} k_i, k_0\}$. Let $\Lambda \in \mathbb{R}^{c\times n}$, where each row of Λ is the concatenation of an image of size $n_x \times n_y$. Let the dictionary Ψ fulfill the A-RIP of order $5k_0\gamma^2$ with constants $\mathcal{L}$ and $\mathcal{U}$. Let $\gamma \geqslant \xi\sqrt{K}/\sqrt{k}$. Furthermore let $\mathcal{A} : \mathbb{R}^{(n_x-1)\times n_y} \to \mathbb{R}^{m_1}$ and $\mathcal{A}' : \mathbb{R}^{(n_y-1)\times n_x} \to \mathbb{R}^{m_1}$ be shifted measurements on the images $\Lambda_i, i = 1,\dots,c$, such that the operator $\mathcal{B} := [\mathcal{A}\ \mathcal{A}',\dots,\mathcal{A}\ \mathcal{A}']$, consisting of c concatenations of $[\mathcal{A}\ \mathcal{A}']$, has the RIP of order $5\sum_{i=1}^{c} k_i\gamma^2$ and level $\delta < 1/3$. Let the operator $\mathcal{D}_{\Phi,\Psi} : \mathbb{R}^{c\times n} \to \mathbb{R}^{m_2\times c}$ possess the D-RIP of order $5k_0\gamma^2$ and level $\delta^* < 1/3$. Consider the linear operator $\mathcal{M} : \mathbb{R}^{c\times n} \to \mathbb{R}^{4cm_1} \times \mathbb{R}^{m_2\times c}$ with components*

$$\mathcal{M}\Lambda = \Big(\mathcal{A}^0\Lambda_1, \mathcal{A}_0\Lambda_1, \mathcal{A}'^0\Lambda_1, \mathcal{A}'_0\Lambda_1, \dots$$
$$\mathcal{A}^0\Lambda_c, \mathcal{A}_0\Lambda_c, \mathcal{A}'^0\Lambda_c, \mathcal{A}'_0\Lambda_c, \mathcal{D}_{\Phi,\Psi}\Lambda \Big). \tag{7.27}$$

If noisy measurements $Y = \mathcal{M}\Lambda + Z$ are observed with noise level $\|Z\|_F \leqslant \varepsilon$, then

$$\Lambda^\diamond = \underset{W\in\mathbb{R}^{c\times n}}{\operatorname{argmin}} \|W\|_1 + \sum_{i=1}^{c} \|W_i\|_{TV} \quad s.t. \quad \|\mathcal{M}W - Y\|_F \leqslant \varepsilon, \tag{7.28}$$

satisfies both

$$\|\Lambda - \Lambda^\diamond\|_F + \sum_{i=1}^{c} \|\nabla\Lambda_i - \nabla\Lambda_i^\diamond\|_F$$
$$\lesssim \frac{1}{\sqrt{K}} \left(\|\Lambda - \Lambda_{S_0}\|_1 + \sum_{i=1}^{c} \left\|\nabla\Lambda_i - (\nabla\Lambda_i)_{S_i}\right\|_1 \right) + \varepsilon, \tag{7.29}$$

and

$$\|\Lambda - \Lambda^\diamond\|_1 + \sum_{i=1}^{c} \|\Lambda_i - \Lambda_i^\diamond\|_{TV}$$
$$\lesssim \|\Lambda - \Lambda_{S_0}\|_1 + \sum_{i=1}^{c} \left\|\nabla\Lambda_i - (\nabla\Lambda_i)_{S_i}\right\|_1 + \sqrt{K}\varepsilon. \tag{7.30}$$

In the formulation of Theorem 7.6.8 the variables k_0 and $\sum_{i=1}^{c} k_i$ appear as the sparsity levels with respect to the full data matrix Λ and its c gradient images $\nabla\Lambda_i$, $i = 1, \ldots, c$, respectively. The restricted condition number ξ as well as the information on the sparsity levels (given via K and k) is controlled by the parameter γ. The latter plays itself an important role in the several orders of the RIP conditions. A detailed discussion on the requirements given is left to the discussion after the proof of the theorem.

Proof of Theorem 7.6.8. For $X \in \mathbb{R}^{c \times n}$ define

$$\|\nabla X\|_{p,\Sigma} := \left(\sum_{i=1}^{c} \|\nabla X_i\|_p^p \right)^{1/p}. \tag{7.31}$$

With respect to Proposition 7.6.6, since $\mathcal{B}$ as well as $\mathcal{D}_{\Phi,\Psi}$ satisfy the RIP, it suffices to show that for $D = \Lambda - \Lambda^\diamond$, both D and $[\nabla D_1, \ldots, \nabla D_c]^T$ satisfy the tube and cone constraints. Write $L_i = [(D_i)_x, (D_i)_y^T]$ and let P denote the map which maps the indices of non-zero entries of ∇D_i to their corresponding indices in L_i. Let $L := [L_1, \ldots, L_c]^T$ and $A_1, A_2, \ldots, A_{m_1}, A_1', A_2', \ldots, A_{m_1}'$ be such that for an image W

$$\mathcal{A}(W)_j = \langle A_j, W \rangle, \quad \text{and} \quad \mathcal{A}'(W)_j = \langle A_j', W \rangle.$$

115

It will now be shown that D as well as $[\nabla D_1, \ldots, \nabla D_c]^T$ satisfy the tube and cone constraints.

Cone constraint: Let S_0 be the support of the s_0 largest entries of Λ, and for $i = 1, \ldots, c$, let S_i denote the support of the s_i largest entries of $\nabla \Lambda_i$ and S_i^c its complement and set $S := \bigcup S_i$. Using the minimality property of $\Lambda^\diamond = \Lambda - D$, it follows that

$$\|\Lambda_{S_0}\|_1 - \|D_{S_0}\|_1 - \|\Lambda_{S_0^c}\|_1 + \|D_{S_0^c}\|_1 +$$
$$\|\nabla \Lambda_S\|_{1,\Sigma} - \|\nabla D_S\|_{1,\Sigma} - \|\nabla \Lambda_{S^c}\|_{1,\Sigma} + \|\nabla D_{S^c}\|_{1,\Sigma}$$
$$\leq \|\Lambda_{S_0} - D_{S_0}\|_1 + \|\Lambda_{S_0^c} - D_{S_0^c}\|_1 +$$
$$\|\nabla \Lambda_S - \nabla D_S\|_{1,\Sigma} + \|\nabla \Lambda_{S^c} - \nabla D_{S^c}\|_{1,\Sigma}$$
$$= \|\Lambda^\diamond\|_1 + \|\nabla \Lambda^\diamond\|_{1,\Sigma}$$
$$= \|\Lambda^\diamond\|_1 + \sum_{i=1}^{c} \|\nabla \Lambda_i^\diamond\|_1$$
$$\leq \|\Lambda\|_1 + \sum_{i=1}^{c} \|\nabla \Lambda_i\|_1$$
$$= \|\Lambda_{S_0}\|_1 + \|\nabla \Lambda_S\|_{1,\Sigma} + \|\Lambda_{S_0^c}\|_1 + \|\nabla \Lambda_{S^c}\|_{1,\Sigma}.$$

Rewriting this inequality leads to

$$\|D_{S_0^c}\|_1 + \|\nabla D_{S^c}\|_{1,\Sigma} \leq \|D_{S_0}\|_1 + 2\|\Lambda_{S_0^c}\|_1 + \|\nabla D_S\|_{1,\Sigma} + 2\|\nabla \Lambda_{S^c}\|_{1,\Sigma}$$
$$= \|D_{S_0}\|_1 + \|\nabla D_S\|_{1,\Sigma} + 2\|\Lambda - \Lambda_{s_0}\|_1 +$$
$$2\|\nabla \Lambda - \nabla \Lambda_S\|_{1,\Sigma},$$

Using the definition of $\| \cdot \|_{p,\Sigma}$ from (7.31) for $p = 1$ yields

$$\|D_{S_0^c}\|_1 + \sum_{i=1}^{c} \|(\nabla D_i)_{S_i^c}\|_1$$
$$\leq \|D_{S_0}\|_1 + \sum_{i=1}^{c} \|(\nabla D_i)_{S_i}\|_1 + 2\|\Lambda - \Lambda_{S_0}\|_1 + 2\sum_{i=1}^{c} \|\nabla \Lambda_i - (\nabla \Lambda_i)_{S_i}\|_1.$$

Now set $\sigma := 2\|\Lambda - \Lambda_{S_0}\|_1 + 2\sum_{i=1}^{c} \|\nabla \Lambda_i - (\nabla \Lambda_i)_{S_i}\|_1$. Using the projection P of the non-zero entries of ∇D_i on each L_i with $|P(S_i)| \leq |S_i|$, it follows that D and L satisfy the cone constraint

$$\|D_{S_0^c}\|_1 + \|L_{P(S)^c}\|_1 \leq \|D_{S_0}\|_1 + \|L_{P(S)}\|_1 + \sigma.$$

Tube constraint: First, D immediately satisfies a tube constraint by feasibility since

$$\|\mathcal{M}D\|_F \leqslant \|\mathcal{M}\Lambda - Y\|_F + \|\mathcal{M}\Lambda^\diamond - Y\|_F \leqslant 2\varepsilon.$$

Using Lemma 4.4.3 from page 64 for the j-th measurement of the derivative in both the x and y directions of each image D_i, $i = 1, \ldots, c$, it follows that

$$\begin{aligned}
|\langle A_j, (D_i)_x \rangle|^2 &= |\langle A_j^0, D_i \rangle - \langle A_{j,0}, D_i \rangle|^2 \\
&\leqslant 2|\langle A_j^0, D_i \rangle|^2 + 2|\langle A_{j,0}, D_i \rangle|^2
\end{aligned}$$

and

$$\begin{aligned}
|\langle A_j', ((D_i)_y)^T \rangle|^2 &= |\langle A_j'^0, (D_i)^T \rangle - \langle A_{j,0}', (D_i)^T \rangle|^2 \\
&\leqslant 2|\langle A_j'^0, (D_i)^T \rangle|^2 + 2|\langle A_{j,0}', (D_i)^T \rangle|^2.
\end{aligned}$$

Thus, L satisfies a tube constraint

$$\|\mathcal{B}L\|_F^2 = \sum_{i=1}^{c} \sum_{j=1}^{m_1} |\langle A_j, (D_i)_x \rangle|^2 + |\langle A_j', ((D_i)_y)^T \rangle|^2 \leqslant 2\|\mathcal{M}D\|_F^2 \leqslant 8\varepsilon^2.$$

To apply Proposition 7.6.6, it remains to show that D also satisfies a tube constraint under the measurements $\mathcal{D}_{\Phi,\Psi}$. But this easily holds since

$$\|\mathcal{D}_{\Phi,\Psi}D\|_F \leqslant \|\mathcal{M}D\|_F \leqslant 2\varepsilon. \qquad \square$$

Remark 7.6.9 (Imposing nonnegativity). A nonnegativity condition can be easily incorporated into Theorem 7.6.8, since only feasibility of the true solution and the minimizer is needed.

Remark 7.6.10 (Main message of Theorem 7.6.8). The robustness result in Theorem 7.6.8 states, under all its technical requirements, that as long as the c gradient measurements given by

$$\left(\mathcal{A}^0 \Lambda_1, \mathcal{A}_0 \Lambda_1, \mathcal{A}'^0 \Lambda_1, \mathcal{A}_0' \Lambda_1, \mathcal{A}^0 \Lambda_c, \mathcal{A}_0 \Lambda_c, \mathcal{A}'^0 \Lambda_c, \mathcal{A}_0' \Lambda_c \right)$$

and the actual measurements on the object $\mathcal{D}_{\Phi,\Psi}\Lambda$ in (7.27) satisfy the appropriate RIP conditions, robust reconstruction via the minimization problem (7.28) is possible. The fact that the individual operators satisfy the respective RIP property depends, as usual, on the given sparsity-level as well as on the nature of the dictionary.

For Theorem 7.6.8 it remains to validate that $\mathcal{D}_{\Phi,\Psi}$ satisfies the D-RIP, $\mathcal{B}$ the RIP and Ψ the A-RIP.

Remark 7.6.11. First note that one can equivalently rewrite the given problem in vectorized form by using the Kronecker product $\otimes$ and the identity [107, Lemma 4.3]

$$(CDE)_{\text{vec}} = (E^T \otimes C)D_{\text{vec}},$$

where the notation $(Z)_{\text{vec}}$ emphasizes the vectorized form of the matrix Z by stacking the columns of Z into a single column vector. With respect to the equations (7.11) and (7.15) it then follows

$$
\begin{aligned}
y := Y_{\text{vec}} &= (I_{c\times c} \otimes \Phi)X_{\text{vec}} \\
&= (I_{c\times c} \otimes \Phi)(\Lambda^T \Psi^T)_{\text{vec}} \\
&= (I_{c\times c} \otimes \Phi)(I_{n\times n}\Lambda^T \Psi^T)_{\text{vec}} \qquad (7.32)\\
&= \underbrace{(I_{c\times c} \otimes \Phi)}_{:=\tilde{\Phi}}\underbrace{(\Psi \otimes I_{n\times n})}_{:=\tilde{\Psi}}\underbrace{(\Lambda^T)_{\text{vec}}}_{:=\tilde{\lambda}}.
\end{aligned}
$$

Therefore, (7.19) or (7.21) can be formulated as

$$
\begin{aligned}
&\underset{\tilde{\lambda}\in\mathbb{R}^{cn}}{\operatorname{argmin}}\|\tilde{\lambda}\|_1 + \sum_{i=1}^{c}\|\Omega_i\tilde{\lambda}\|_{TV} \\
&\text{subject to } \|y - \tilde{\Phi}\tilde{\Psi}\tilde{\lambda}\|_2 \leq \varepsilon,\ \tilde{\lambda} \geq 0,
\end{aligned}
\qquad (7.33)
$$

where Ω_i maps the i-th m/z-image in the vector $\tilde{\lambda}$, indexed by $[(i-1)n+1, in]$, to its corresponding $n_x \times n_y$ matrix form. The optimization problem in (7.33) has now the typical compressed sensing form where the sensing matrix Φ is multiplied with the basis matrix first, followed by the unknown coefficient vector $\tilde{\lambda}$ (cf. (4.17), page 52 and (4.23), page 54). The resulting matrix $\tilde{\Phi}$ is a $mc \times nc$ blockdiagonal matrix with entries Φ on the diagonal.

Remark 7.6.12 (Assumptions of Theorem 7.6.8). If one can now show that the D-RIP holds for $\tilde{\Phi}$, it also follows that $\mathcal{D}_{\Phi,\Psi}$ fulfills the D-RIP, since

$$\|\Lambda^T\Psi^T\|_F^2 = \|\Psi\Lambda\|_F^2 = \|\tilde{\Psi}\tilde{\lambda}\|_2^2 \quad \text{and} \quad \|\tilde{\Phi}\tilde{\Psi}\tilde{\lambda}\|_2^2 = \|\mathcal{D}_{\Phi,\Psi}\Lambda\|_F^2.$$

In [84] it has very recently been shown that the RIP (or also the D-RIP) holds with overwhelming probability also for blockdiagonal matrices if the elements of the matrix Φ are independently drawn at random from subgaussian distributions, the underlying basis Ψ is an orthogonal basis and if

$$mc \gtrsim Kc \log^2(nc) \log^2(K), \qquad (7.34)$$

where $K = \max\{\sum_{i=1}^{c} k_i, k_0\}$ is the maximal sparsity level.

With respect to the underlying system and its parameters, the only gap in the result (7.34) as to the applicability to the given setting lies in the requirement to have an orthogonal matrix. Here, Ψ is not orthogonal but close to one if the standard deviation σ in the Gaussian (7.14) would be set very small. In our case, this result from [84] is only an approximate value for the number of measurements m needed to fulfill the RIP. Moreover, due to the quadratic log-expressions in (7.34) the number m is far to large to get reasonable values. Put differently, the IMS data would need to be very sparse to get a small number of measurements required. Also the authors from [84] state that the relation (7.34) may not be optimal and can be improved most likely. In fact, as just discussed for the given IMS problem, the estimate for m in (7.34) is way to restrictive if one takes a priori information about the problem into account.

Concerning the A-RIP in Theorem 7.6.8, note that the used dictionary $\Psi \in \mathbb{R}_+^{c \times c}$ is invertible and therefore its condition number κ provides an upper bound for all ξ. Starting with the identity $Z = \Psi^{-1}\Psi Z$ for a matrix $Z \in \mathbb{R}^{c \times n}$ it is $\|Z\|_F \leqslant \|\Psi^{-1}\|_F \|\Psi Z\|_F$. Thus, it is

$$\frac{1}{\|\Psi^{-1}\|_F} \|Z\|_F \leqslant \|\Psi Z\|_F \leqslant \|\Psi\|_F \|Z\|_F,$$

and ξ is obviously bounded by

$$\xi = \frac{\mathcal{U}(\Psi)}{\mathcal{L}(\Psi)} \leqslant \kappa := \|\Psi^{-1}\|_F \|\Psi\|_F.$$

Therefore, Ψ fulfills the A-RIP and ξ from Theorem 7.6.8 can be estimated from above by κ. In the given case, the condition number is small if σ from (7.14) is small, e.g. for $\sigma = 0.75$. As the later numerical results in Section 7.7 will show, this value is a reasonable assumption.

In Theorem 7.6.8 it is required that

$$\gamma \geq \xi\sqrt{K}/\sqrt{k}.$$

If γ, as a RIP order parameter, would be too big it would potentially require more measurements for the underlying measurement matrix to fulfill the RIP property. The value ξ has already been discussed before, so $\sqrt{K}/\sqrt{k}$ needs also to be small. Since Λ can be interpreted as the deconvoluted version of the datacube X (see Figure 7.1 and (7.15)), it inherents the same physical structures. It is therefore reasonable to assume that the sparsity prior $\sum_{i=1}^{c} k_i \approx k_0$ holds for Λ as well and this implies $\sqrt{K} \approx \sqrt{k}$.

Regarding the RIP of $\mathcal{B}$, one could transform the operator into four blockdiagonal matrices as follows:

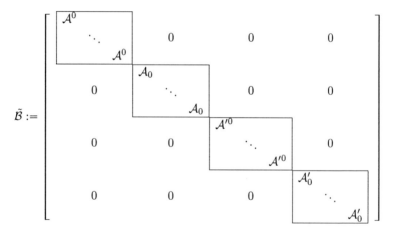

Note that $\tilde{\mathcal{B}}$ performs the $4cm_1$ measurements on the gradients given in the operator $\mathcal{M}$ in Theorem 7.6.8. Now, a discussion about the RIP on this blockdiagonal matrix can be done as before.

Remark 7.6.13 (Gradient measurements). The gradient measurements in Theorem 7.6.8 could theoretically be obtained by shifting the measurement mask. However, in the acquisition process of MALDI-TOF, an ionization of the biological sample is performed. Therefore, the sample is damaged at the ionized points and further measurements would make no sense. As the authors of [151] it is reasonable to believe that the additional $4m_1$ measurements $\mathcal{A}^0, \mathcal{A}_0, \mathcal{A}'^0, \mathcal{A}'_0$ in Theorem 7.6.8 are not necessary. Indeed the numerical results in following Section 7.7 seem to confirm this.

7.7. Numerical results

In this section reconstruction results for an example IMS dataset based on the proposed model are presented. The numerical minimization of the proposed optimization problem (7.21) is done via the parallel proximal splitting algorithm (PPXA) [59], cf. Section 2.2. To improve the ℓ_1 as well as the TV minimization effects, additional regularization parameters $\alpha, \beta > 0$ are introduced. Thus, the optimization problem becomes

$$\underset{\Lambda \in \mathbb{R}^{c \times n}}{\text{argmin}} \ \alpha \|\Lambda\|_1 + \beta \sum_{i=1}^{c} \|\Lambda_i\|_{TV} \tag{7.35}$$
$$\text{subject to} \ \|Y - \mathcal{D}_{\Phi,\Psi}\Lambda\|_F \leqslant \varepsilon, \ \Lambda \geqslant 0.$$

For the PPXA, this is rewritten as a sum of four lower semicontinous convex functions

$$\underset{\Lambda \in \mathbb{R}^{c \times n}}{\text{argmin}} \ f_1(\Lambda) + f_2(\Lambda) + f_3(\Lambda) + f_4(\Lambda), \tag{7.36}$$

where $f_1(\Lambda) = \alpha\|\Lambda\|_1, f_2(\Lambda) = \beta \sum_{i=1}^{c} \|\Lambda_i\|_{TV}, \ f_3(\Lambda) = \iota_{\mathcal{B}_2^\varepsilon}(\Lambda)$ and $f_4(\Lambda) = \iota_{\mathbb{R}_+^{c \times n}}(\Lambda)$. There, ι_C denotes the indicator function which is applied to the convex sets $\mathcal{B}_2^\varepsilon, \mathbb{R}_+^{c \times n} \subset \mathbb{R}^{c \times n}$, corresponding to the matrices that satisfy the fidelity constraint $\|Y - \mathcal{D}_{\Phi,\Psi}\Lambda\|_F \leqslant \varepsilon$ and to the ones lying in the positive orthant, respectively.

The well-studied dataset $X \in \mathbb{R}_+^{n \times c}$ is acquired from a rat brain coronal section (cf. Figure 5.2, page 78). For testing purposes, a smaller part of the full mass range is selected which consists of $c = 2000$ data bins ranging from $4213\,\mathrm{Da}$ to $9104\,\mathrm{Da}$ with spatial dimension $n = 24{,}120$. For visualization, the spectra are normalized using total ion count (TIC) normalization, which is the normalization with respect to the ℓ_1-norm [67].

In the following experiments, the mass spectra are assumed to be sparse or compressible with respect to shifted Gaussians as in (7.14), where the standard deviation is set as $\sigma = 0.75$. By this, the idea of the peak picking as well as a low conditioning number $\xi \approx 8$ will be kept, since the last is an important factor in the robustness Theorem 7.6.8.

The measurement matrix Φ is randomly filled with numbers from an i.i.d. standard normal Gaussian distribution.

The initial guess Λ_0 for the desired solution Λ was set as a random matrix whose negative elements were set to zero.

By experience, the noise level ε was set to 3.75×10^3 and there were 30 outer loop iterations applied in the PPXA. The regularization parameters in (7.35) were set for each amount of measurements by hand as follows: 20%: $\alpha = 0.15$, $\beta = 0.3$, 40%: $\alpha = 1.3$, $\beta = 1.6$, 60%: $\alpha = 2.0$, $\beta = 2.3$, 80%: $\alpha = 3.2$, $\beta = 3.5$ and 100%: $\alpha = 4.8$, $\beta = 5.1$.

Figure 7.4 presents the mean spectrum, i.e. the sum over all pixel spectra $X_{(i,\cdot)}$ for $i = 1, \ldots, n$, of the rat brain data as well as the mean spectrum of the reconstruction, based on 20%, 40% and 60% measurements taken. The triangles in Figure 7.4(a) are selected manually and show the peaks which are detected based the 20% level. Figures 7.4(b) and 7.4(c) show which peaks are additionally extracted during reconstruction. They are, as for the triangles, manually selected and the peaks are visualized by additional squares and circles. Clearly visible is the influence of more measurements on the feature extraction of the main peaks in the mean spectrum. As an example, the peak at 7060 Da is only slightly extracted in 7.4(a). More measurements not only lead to a higher intensity of this peak, but also in additional local information, see Figure 7.4(b)-(c). Note that the described effect is only caused by the amount of samples and does only slightly alter with the regularization parameters.

The effect of this increasing peak intensities can be visualized, for example, by looking at the corresponding m/z-image at m/z 7060, see Figure 7.6. At the 20% level the main spatial features of this image are extracted, but details such as in the lower portion of the data are missed. Increasing the number of measurements clearly leads to better reconstruction results. With 40% of taken measurements one gets almost all main features of the m/z-image.

Finally, Figures 7.7 and 7.8 show images for six additional m/z-values and their corresponding reconstructions at the different measurement lev-

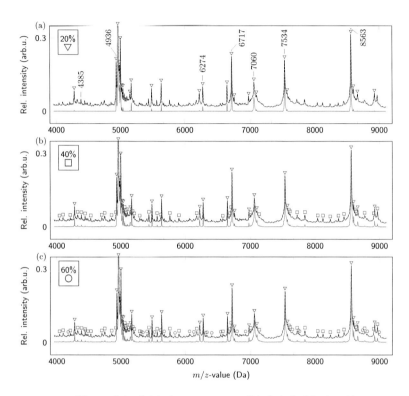

Figure 7.4.: Original mean spectrum (black dashed line) and its reconstruction based on different number of taken measurements (blue line). The spectra are both normalized to $[0, 1]$ and the upper half is leaving out for better visualization. (a) Reconstruction based on 20% out of $n = 24{,}120$ taken compressed measurements. The triangles ($\triangledown$) are set manually and visualize which peaks are mainly detected. (b) and (c) show reconstruction results for 40% and 60%, respectively. The squares ($\square$) and circles ($\bigcirc$) are also manually selected and show which peaks appear to be additionally detected.

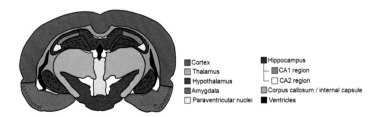

Figure 7.5.: Schematic representation based on the rat brain atlas. Reprinted with permission from [9]. Copyright 2010 American Chemical Society.

els (20%, 40%, 60%, 80% and 100%). These m/z-values correspond to six detected high intensity peaks in the mean spectrum as visualized in Figure 7.4(a). Moreover, the m/z-images present main structures within the rat brain, as shown in the rat brain schematic in Figure 7.5, adapted from [9]. As one can see, regions of high intensity pixels are mostly detected as such and were reconstructed well when using 40% measurements. In Figures 7.7 and 7.8, one also notices a slight loss of details when applying 40% measurements, as seen previously in Figure 7.6. This loss clearly reduces with the amount of measurements taken. The image at 4385 Da illustrates the reconstruction results on a smaller peak, compared to the other selected. It is recognizable that 20% taken measurements lead to only an idea of where regions appear in the measured image sample, see also 7.4(a) in comparison with 7.4(b). In contrast, 40% taken measurements lead to reasonable reconstruction results. This behavior can be observed on the other reconstructed m/z-images as well.

It needs to be mentioned that there were no additional measurements on the gradients taken as they are required in Theorem 7.6.8. As it is mentioned in Section 7.6, it is reasonable to believe that they are not required in practice. Moreover, the actual theoretical bound in (7.34) on the number of measurements seems to be too pessimistic. For the rat brain example only few (around 40%) are needed for good reconstruction results.

In Appendix A, similar results are presented for the rat kidney data which was described in Section 5.3.2. There, a reconstruction based on only 20% could not be found for all set of regularization parameters tested. This is probably a result that can be interpreted using the in-

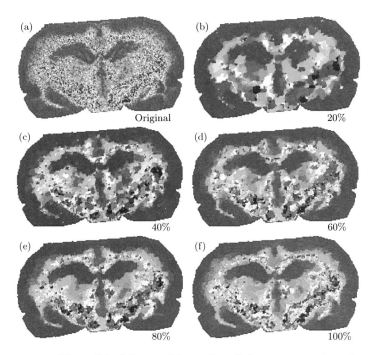

Figure 7.6.: Influence of the number of taken measurements on the reconstruction of the full dataset; an m/z-image corresponding to 7060 Da is shown. (a) shows the original image. The images in (b), (c), (d), (e) and (f) show the reconstructions with respect to 20%, 40%, 60%, 80% and 100% taken measurements, respectively.

equality (7.34). Compared to the rat brain dataset where there were given $c = 2000$ channels, the rat kidney data comprises of $c = 10{,}000$ channels. The parameter c appears in the inequality on the right hand side in a quadratic log-term. The relation states that more measurements m are needed to guaranty that the measurement matrix fulfills the RIP and thus, robust recovery of the data is possible.

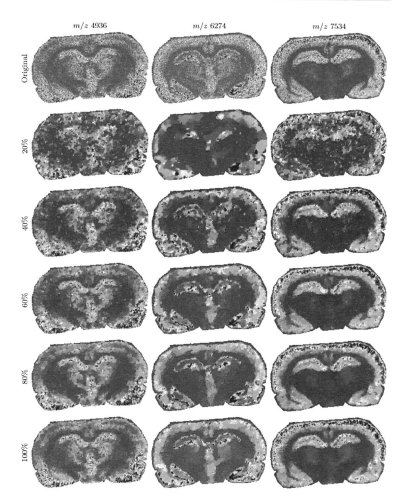

Figure 7.7.: Reconstructions of three different m/z-images based on 20%, 40%, 60%, 80% and 100% of taken measurements. First column, 4936 Da with main structures in the middle and the lower part; second column, 6274 Da with structures at the boundaries and small regions of high intensity pixels in the middle and bottom part; third column, 7534 Da with high intensities at the boundary as well as in the middle and the bottom.

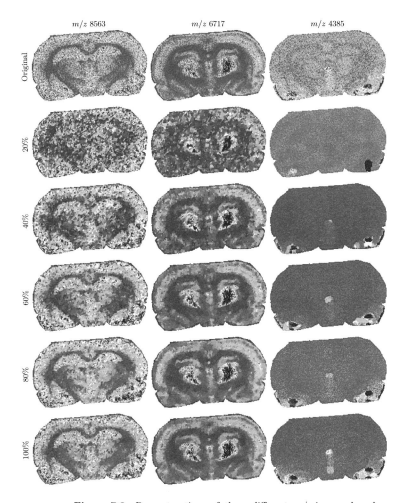

Figure 7.8.: Reconstructions of three different m/z-images based on 20%, 40%, 60%, 80% and 100% of taken measurements. First column, 8563 Da with structures at the boundary as well as in the middle and the lower part; second column, 6717 Da with one main structure in the center and less intensive regions at the boundary; third column, 4385 Da with only small spots of high intensity pixels in the middle and the very bottom.

7.8. Conclusion

In this chapter, a first model for CS in IMS was investigated. In addition, it was theoretically proven that both the reconstruction of the spectra and the m/z-images is robust. The results presented here except the rat kidney in Appendix A were published in [18, 19].

Currently there are no mass spectrometers which allow for the acquisition of data in such manner. However, considering the recent developments of the single pixel camera [79, 183], one could theoretically implement such a mass spectrometry by splitting the laser into several beams analogously as it is done in the digital micromirror device used in the single pixel camera. Then, instead of analyzing each pixel separately, one could analyze several pixels simultaneously and accumulate a measurement-mean spectrum for such a measurement. Note that modern mass spectrometers indeed use complex optics to achieve non-flat structured laser profile as in Bruker Daltonics smartbeam mass spectrometers [106], although the current optics does not allow to change the profile during an experiment.

The mathematical model for hyperspectral IMS data covers the two typical post-processing steps of peak picking of the spectra and denoising of the m/z-images and applies the compression of the hyperspectral IMS datacube during the measurement process. The reconstructed images as well as the spectra were shown to capture the features both in the spatial and the spectral domains. As visually judged, taking 40% to 60% of the typical number of measurements led to only a slight loss of spatial features even of small size. Recently, similar numerical results were designed for an desorption electrospray ionization (DESI) mass spectrometer [95]. There, the authors also state that they are able to halve the measurement time, but did not give any theoretical results.

The presented robustness result given in Theorem 7.6.8 combines the ℓ_1- and the TV-results described in Chapter 4. For this, both sparsity aspects in spectral and spatial dimension needed to be considered. Moreover, the general setup is multidimensional which is so far covered for TV only by Needell and Ward [150]. In addition, the theorem is not restricted to orthonormal basis. Instead, any frame or overcomplete dictionary can be taken. In case of an invertible dictionary, the condition number will have an impact on the required number of measurements m. For the latter, a RIP result on block-diagonal matrices for orthogonal matrices from [84] has been used to give an estimate for m for different system parameters. What remains open is the extension of this block-diagonal result to non-orthogonal matrices.

The parallel proximal algorithm [59] was used to solve the optimization problem. To improve the regularization effects, regularization parameters α and β were added and set by hand for each different amount of measurements. As it can be slightly seen in the results (e.g. 20% in Figure 7.7), it is not feasible to set α by hand for all images. A future direction of investigation should therefore involve regularization terms with locally-dependent parameters $\alpha_i = \alpha(X_i)$ for $i = 1, \ldots, c$ for the m/z-images as in [9] for locally-adaptive denoising, and $\beta_j = \beta(X_{(j,\cdot)})$ for $j = 1, \ldots, n$ for the spectra. Furthermore, one could study other algorithms such that in [127] for improving the calculation speed from actually several hours to only few minutes.

A possible generalization of the presented CS in IMS model is to combine it with the discussed NMF, cf. Sections 3.3.3 and 6.4. On that way, one would not reconstruct the full (feature) dataset $\Lambda \in \mathbb{R}_+^{c \times n}$, but extract its $\rho \ll \min\{n, c\}$ most characteristic fingerprint spectra and soft segmentation maps. Formally, one would aim to reconstruct two much smaller matrices $P \in \mathbb{R}_+^{c \times \rho}$ and $S \in \mathbb{R}_+^{\rho \times n}$, such that

$$\Lambda \approx PS.$$

The resulting optimization problem then reads

$$\operatorname*{argmin}_{P \in \mathbb{R}_+^{c \times \rho}, S \in \mathbb{R}_+^{\rho \times n}} \|P\|_1 + \sum_{i=1}^{\rho} \|S_i\|_{TV} \tag{7.37}$$
$$\text{subject to } \|Y - \mathcal{D}_{\Phi,\Psi} PS\|_F \leqslant \varepsilon, \ \Lambda \geqslant 0.$$

For (7.37), next to a numerical analysis, one could study if it is possible to give a similar robustness result as given in this chapter.

At last, future work might also replace the Gaussian noise model with a Poisson statistics approach [158, 165] or with a combination of both as observed in [12]. As it has been mentioned in [9, 115], this model might be more suitable for MALDI-TOF mass spectrometry.

8 | Compressed sensing based multi-user detection

8.1. Introduction

This chapter addresses the application of CS based multi-user detection (MUD) in the field of wireless sporadic communication on the example of the so-called Code Division Multiple Access (CDMA) system. At first, a known model for a CDMA system is presented, followed by a brief explanation how it is connected to CS. In addition, a new parameter identification principle that follows the idea of the L-curve [103] for a CDMA setting that uses the elastic-net functional for reconstruction of the transmitted signal is introduced. The presentation of the CDMA system is based on [30] whereas the parameter identification approach was developed in cooperation with the Department of Communications Engineering (ANT) of the Institute of Telecommunications and High-Frequency Technique (ITH) from the University of Bremen and published in [167].

8.2. Sporadic communication

The term *sporadic communication* refers to a scenario in which information exchange between certain points takes place only irregularly and seldom. It is a special case of the general machine-to-machine (M2M) communication which summarizes all technologies that enable the communication between technical devices (e.g. robot-to-robot) without human intervention [204]. Because of the rising numbers of users in different M2M systems, one expects a growing interest in M2M communication

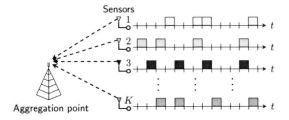

Figure 8.1.: Model for the sporadic communication scenario. K sensor nodes are able to communicate with a central aggregation point, but it is assumed that only a very few are active at the same time.

in the next years [20, 61]. Most of the M2M scenarios can be divided in two cases. One is the *star topology* in which multiple sensor nodes (*multi-user*) communicate with a central aggregation, as illustrated in Figure 8.1. The other is the *mesh network* describing a system of nodes that are additionally able to communicate with each other.

In this chapter, a special case of the first model is considered in which all sensors follow a *multiple access* control to share their propagation channels (or in short *channel*) and to communicate over them. Moreover, the sensor nodes are only occasionally active, which leads to the field of *sporadic communication* [30]. While the sensors themselves are assumed to have a low complexity, the aggregation point, however, has the ability to do complex signal processing.

As every node of a multi-user system is able to transmit at the same time, it is reasonable to ask how the aggregation point can reconstruct both the activity and the transmitted data. Traditional approaches for this task are for example given by Time Division Multiple Access (TDMA) and Frequency Division Multiple Access (FDMA) where the physical channel is divided into time and frequency resources, respectively. More specifically, in this context one speaks of *orthogonal* resources, since in TDMA and FDMA the data is transmitted orthogonally. Both have the problem that they reserve bandwidth for transmission to distinguish between users, regardless of whether they are currently being used or not. The so-called Code Division Multiple Access (CDMA) transmits user information in a code space spanned over time and frequency.

In this case, the physical channel is given by a possible non-orthogonal *code* resource, where a code is a set of symbols describing a certain information. CDMA is used in, for example, the current Universal Mobile Telecommunications System (UMTS) [193]. Since CDMA is the example setup in this chapter for multiple-user and sporadic communication in wireless systems, a describing model will be introduced hereafter.

8.3. Multi-user system modelling

A multi-user communication setup, as it is illustrated in Figure 8.1, can formally described as a system in which each of the K nodes ($1 \leqslant k \leqslant K$) aims to send a sequence of L data packets or *symbols* $x_{k,\ell}$, $1 \leqslant \ell \leqslant L$, to the same destination. The information message of node k is collected in the symbol vector

$$x_k = (x_{k,1} \ \ldots \ x_{k,L})^T \in \mathbb{R}^L,$$

where each symbol $x_{k,\ell}$ is taken from a pre-specified and system dependend symbol alphabet $\mathcal{A}$. In CDMA, the symbols $x_{k,\ell}$ are not transmitted directly to the aggregation point, but are spread by a node-specified pseudo-random spreading sequence $s_k \in \mathcal{S}^N$ of length N with elements from a set $\mathcal{S}$. This leads to the so-called *chip* vector $c_{k,\ell} = s_k x_{k,\ell}$. On the one side, depending on the predefined length N, a larger bandwidth is needed. But on the other side, this enables to use the transmission channel concurrently for a larger number of sensors. In the sum, the k-th node sequently transmits in total $F = LN$ chips over the communication channel. This is done at the Nyquist rate [152] which gives an upper bound for the symbol rate for a band-limited channel for omitting aliasing or, in other words, to fulfill the Shannon-Nyquist sampling theorem [175]. In matrix form the whole chip vector for the k-th node reads

$$c_k = S_k x_k, \tag{8.1}$$

where S_k denotes the $F \times L$ node-specific blockdiagonal spreading matrix

$$S_k = \begin{pmatrix} s_k & 0 & \cdots & 0 \\ 0 & s_k & & 0 \\ \vdots & & \ddots & \\ 0 & 0 & & s_k \end{pmatrix}.$$

133

As it is mentioned in [198], "wireless communication systems are affected by propagation anomalies due to terrain or buildings which cause multipath reception, producing extreme variations in both amplitude and apparent frequency in the received signals, a phenomenon known as *fading*." This can be modeled as follows: The channel from the k-th node to the aggregation point is identified by its channel impulse response $h_k \in \mathbb{R}^{N_H}$ and the influence on the transmitted chips is computed by the convolution with h_k [30]. Usually, each node experiences a different physical channel to the receiver and different impulse responses may contain different numbers of coefficients. For simplicity, it is assumed that all channel impulse responses contain the same number of N_H fading coefficients. Due to the convolution with the channel responses $h_k \in \mathbb{R}^{N_H}$, which can be described by the multiplication of the chip vector c_k from (8.1) with the following $F' \times F$ Toeplitz structured convolution matrix

$$
H_k = \begin{pmatrix} h_{k,0} & 0 & \cdots \\ h_{k,1} & h_{k,0} & \\ \vdots & \vdots & \ddots \\ h_{k,N_H-1} & h_{k,N_H-2} & \\ 0 & h_{k,N_H-1} & \\ 0 & 0 & \ddots \end{pmatrix},
\tag{8.2}
$$

the F transmitted chips of one node result in $F' = F + N_H - 1$ received chips at the destination. Under der assumption that all nodes start the transmission of the sequence of chips c_k simultaneously, the overall received signal vector $y \in \mathbb{R}^{F'}$ at the aggregation point is given by the linear superposition of all K transmitted vectors c_k, convolved with the corresponding channel impulse response h_k.

By including also a white Gaussian distributed noise vector $n \in \mathbb{R}^{F'}$ with zero mean and variance σ_n^2 to the process, i.e. $n_i \sim \mathcal{N}(0, \sigma_n^2)$, $i = 1, \ldots, F'$, it follows

$$
y = \sum_{k=1}^{K} H_k c_k + n = \sum_{k=1}^{K} H_k S_k x_k + n.
\tag{8.3}
$$

Note at this point that for incorporated noise $y = x + n$ it is common in engineering to express noise $n(t)$ in terms of signal to noise ratios (SNR) (see also Section 3.2) in decibels (SNR$_{dB}$) by[1] (cf. [112])

$$\text{SNR}_{dB} = 10\log_{10}(\text{SNR}) = 10\log_{10}\left(\frac{\mathbb{E}\{|x(t)|^2\}}{\mathbb{E}\{|n(t)|^2\}}\right). \qquad (8.4)$$

The numerator and the denominator in (8.4) both describe the second moments. If a random variable z is $\mathcal{N}(\mu, \sigma^2)$-distributed, the second moment becomes $\mathbb{E}\{|z(t)|^2\} = \mu^2 + \sigma^2$. In the following, the notation SNR will be synonymously used for the SNR in decibel as expressed in (8.4).

By collecting the convolution and the spreading matrices of all nodes into matrices $H \in \mathbb{R}^{KF' \times KF}$ and $S \in \{-1, 0, 1\}^{KF \times KL}$,

$$H = \begin{pmatrix} H_1 & 0 & \cdots & 0 \\ 0 & H_2 & & 0 \\ \vdots & & \ddots & \\ 0 & 0 & & H_K \end{pmatrix} \quad \text{and} \quad S = \begin{pmatrix} S_1 & 0 & \cdots & 0 \\ 0 & S_2 & & 0 \\ \vdots & & \ddots & \\ 0 & 0 & & S_K \end{pmatrix},$$

and by defining the $F' \times KF'$ matrix

$$M = \begin{pmatrix} I & I & I & \cdots \end{pmatrix} \qquad (8.5)$$

containing K identity matrices $I \in \mathbb{R}^{F' \times F'}$, the input-output relationship in (8.3) can be written as

$$y = Ax + n, \qquad (8.6)$$

with the $F' \times KL$ system matrix

$$A = MHS. \qquad (8.7)$$

The concatenated symbol vector or *signal* $x = (x_1 \ldots x_K)^T \in \mathcal{A}^{KL}$ contains all KL information symbols of the K nodes. The matrix A executes the linear superposition of the distorted signals of all nodes, H represents the impact of the physical channels (so-called *distortion*) to the transmission and S is the spreading matrix with all spreading sequences of all nodes.

---

[1]The expression $\mathbb{E}\{\cdot\}$ denotes the *expected value* for a function $f(x)$ with random variable x and is given as $\mathbb{E}\{f(x)\} = \int_{-\infty}^{\infty} f(z)p_x(z)\,dz$ where p_x is a probability density function.

Remark 8.3.1. In the communications community, one often reads the terms *compressed sensing multi-user detection* (CS MUD) and *compressed sensing based multi-user detection*, see e.g. [30, 94, 168–170, 176, 208]. This is due to the fact that the taken measurements y, taken by the matrix A, consist of a *random* mixture of the transmitted user signals. The randomness arises at first because of the node-specific random spreading sequences s_k. In addition, the physical channels are often statistically modeled as so-called *Rayleigh fading* [112]. This model assumes to have a large number of scattering objects and that there is no direct transmission path. Thus, the amplitude of the transmitted signal will vary at random following the Gaussian distribution. However, what is not realized by the above model is a CS typical undersampling of the transmitted signals.

In connection with Remark 8.3.1, it should also be described the difference from the usual CS measurement matrices shape (cf. (4.17) on page 52) $A = \Phi\Psi$ to the matrix A in (8.7). As it is well described in [30], "... in contrast to the typical CS problem, the measurement matrix Φ cannot be chosen arbitrarily, but is given physically, and the sparsity basis Ψ is not constrained by physical properties of a signal, but can be chosen freely. Instead of viewing sparsity basis Ψ and measurement matrix Φ as separate parts of the system description, it is also widely accepted to combine the two matrices to an overall matrix A that characterises the transition from the sparse domain to the observation (...), which is exactly the CDMA system Equation (8.6)." Put differently, in light of (8.7) the matrix A can be interpreted as the measurement Φ of the signal x and the node-specific spreading matrix S as a sparse transformation Ψ. Only H in (8.7) does not match the typical CS setup as it models a distortion, i.e. the impact of the physical channels, during transmission that is normally not considered. As in [167] H will be considered to be known, but an analysis of its influence remains open.

Nevertheless, given the relation (8.6), one faces the problem of reconstructing the transmitted signal x from noisy measurements y. The rest of this chapter examines how this can be done via the elastic-net functional and presents a parameter choice rule which is motivated from the known L-curve strategy.

8.4. The elastic-net

It is observable that the system matrix $A \in \mathbb{R}^{F' \times KL}$ has less rows than columns as long as the spreading factor N is chosen smaller than the number of nodes K and if the number of fading coefficients N_H (cf. (8.2)) is reasonable small. Such a scenario is typically referred as *overloaded* CDMA[2] which is used in "applications in which the band is a limited resource and more users are to conveyed on the same channel" [162]. Hence, the linear system (8.6) is highly underdetermined, meaning that it is hopeless to reconstruct the transmitted signal $x \in \mathbb{R}^{KL}$ from taken measurements $y \in \mathbb{R}^{F'}$ without any prior information. Moreover, if one takes into account the assumption that the *active probability* is very low for all nodes, the overall sent signal x is supposed to be highly sparse. This directly motivates (see Chapter 4) to solve either a minimization problem as in (4.23) on page 54, namely

$$\operatorname*{argmin}_{z \in \mathbb{R}^{KL}} \|z\|_1 \text{ subject to } \|Az - y\|_2 \leq \varepsilon,$$

or to minimize the related ℓ_1-penalized Tikhonov functional, see e.g. [66, 126],

$$T_\alpha^1(x) := \frac{1}{2}\|Ax - y\|_2^2 + \alpha\|x\|_1, \tag{8.8}$$

with some regularization parameter $\alpha \geq 0$. Note that the functional T_α^1 in (8.8) is often named Basis Pursuit Denoising (BPDN) in the CS literature, cf. [56, 93] and the references therein. Anyway, there exist many different minimization algorithms for the minimization of this functional, e.g. the iterative soft-thresholding [66], the GPSR algorithm [90] and miscellaneous active set methods, see [57, 102, 122]. However, this yields to inverting ill-conditioned operators, as it is described in for example [109]. Moreover, as discussed in detail by Loris [128], minimization of the pure ℓ_1-functional (8.8) is potentially very slow due to the bad convexity and smoothness properties of the functional T_α^1 which causes problems in finding effective minimization algorithms.

---

[2]The case $N = K$ is named *full-loaded* CDMA in which it is a usual approach to use the pseudoinverse for reconstructing the transmitted signal x in $y = Ax$. The case left is $N < K$ which leads to *underloaded* CDMA. [162]

To counteract this problem, it has been found to be useful to enhance the functional by adding a small ℓ_2-term, leading to the so-called *elastic-net functional* which has been introduced in [209]. It combines ℓ_1-penalized Tikhonov regularization T_α^1 (8.8) and the classical ℓ_2-penalized Tikhonov functional [187, 188], and is defined by

$$T_{\alpha,\beta}^{\text{EN}}(x) := \frac{1}{2}\|Ax - y\|_2^2 + \alpha\|x\|_1 + \beta\|x\|_2^2, \tag{8.9}$$

where $\alpha, \beta \geqslant 0$ are the regularization parameters. A unique minimizer is guaranteed if either $\beta > 0$ or if the operator A is injective, as the elastic-net functional $T_{\alpha,\beta}^{\text{EN}}$ then becomes strictly convex, see e.g. [172, Lemma 4.2.1 and 4.2.2]. Nevertheless, the additional parameter β is fixed for improving the condition of A [109, 209]. It can be set system independent, but should be chosen strict positive but as small as possible, as it is clear that the bigger β the bigger the difference between the ℓ_1 minimizer of (8.8) and the elastic-net minimizer of (8.9). In this chapter, the elastic-net functional (8.9) will be used with a fixed $\beta = 10^{-10}$ for improving the condition of A, while at the same time not destroying the desired sparsity [109].

Remark 8.4.1. In contrast to the parameter β, the choice of $\alpha > 0$ is crucial to the performance of the reconstruction. An optimal choice would require the knowledge of x. Therefore, the main task is to choose a good (and in best case almost optimal) α without the knowledge of x but with the (feasible) knowledge of the SNR and the properties of A. In the considered sporadic communication scenario, both the transmitted signal x as well as the noise level may change over time, such that α has to be tuned continuously to achieve the best performance [167].

To solve the elastic-net functional (8.9) the regularized feature-sign-search (RFSS) algorithm is used, which is proven to converge globally in finitely many steps, see [109] and [172, Theorem 4.4.6]. In a feature-sign-search algorithm, one creates an *active set* of all nonzero coefficients with their corresponding signs and searches systematically for the optimal active set [122]. The RFSS mainly updates the searched-for solution x in each iteration by a regularized variant. For a detailed presentation and discussion refer [172] and the references therein.

8.5. The multi-user test setup

Within this chapter, the same assumptions as in [167] and [168] are made: The underlying system is supposed to have $K = 128$ nodes, each sending $L = 1$ symbol from the so-called *augmented* alphabet $\mathcal{A}_0 = \mathcal{A} \cup \{0\} = \{-1, 0, 1\}$, where $\mathcal{A}$ is here the *finite* modulation alphabet describing the called *Binary Phase Shift Keying* (BSPK), i.e. $\mathcal{A} = \{-1, 1\}$ [112, 167, 169]. The zero symbol indicates inactivity or equivalently that no data is sent. The spreading sequences $s_k \in \mathcal{S}^N$ are ℓ_2-normalized, have length $N = 32$, are taken from the set $\mathcal{S} = \{-1, 1\}$ and are node specific pseudo-random noise (PN) sequences that spread each of the information symbols $s_{k,\ell}$. Due to the identity $F = LN$ it then also follows that each node transmits $F = 32$ chips to the aggregation point. Since it is known that physical channel properties are well described by stochastically models [112], the element in the vectors $h_k \in \mathbb{R}^{N_H}$ are assumed to be $N_H = 6$ i.i.d. Gaussian distributed taps with an exponential decaying power delay profile [168]. Therefore, the system matrix A in (8.7) is of size 37×128, since $F' = F + N_H - 1 = 37$ and $KL = 128$. The activity of each node x_k is modeled by an activity probability p_a which is expected to be the same for all. The main assumption is that most of the time the majority of the K sensors is silent which implies that p_a is rather small, i.e. $p_a \ll 1$. Here, an active probability of $p_a = 0.02$ is assumed.

At last, it needs to be mentioned that the elements of the reconstructed signal are restricted to be elements of the finite augmented BPSK alphabet since $\mathcal{A}_0 = \{-1, 0, 1\}$. This means that the reconstruction of the transmitted signal x via the RFSS algorithm follows a quantization step to the quantized vector $\hat{x} \in \mathcal{A}_0^{KL}$ by

$$\hat{x}_i = \begin{cases} \text{sign}(x_i) & \text{for } |x_i| > \tau_i \\ 0 & \text{otherwise} \end{cases}, \tag{8.10}$$

where $\text{sign}(\cdot)$ is the signum function as defined in (2.13) on page 22 and the threshold τ_i is set as $\tau_i = 0.5\|A_i\|_2$. The factor 0.5 appears due to the assumption to have an equal distribution of the symbols $-1, 0, 1$ and a correct scaling of the non-quantized solution x. The additional factor $\|A_i\|_2$ comes into play since the overall system matrix A will in general not have a unit-norm columns to due the physical channel properties. Therefore, a normed version of A is considered which leads to an amplification of each component of x by its corresponding column norm.

8.6. A parameter choice rule: The C-curve criterion

As described in Section 8.4, the elastic-net functional $T_{\alpha,\beta}^{\text{EN}}$ in (8.9) will be minimized for reconstructing the transmitted signal x. While the parameter β is fixed and can be set system independent, the choice of α, however, has a crucial influence on the expected sparsity of the reconstructed x. In this section a new a posteriori or *online* parameter choice rule for the regularization parameter α is described, which will in the following be called the *C-curve* method. The term 'online' is motivated from the fact that this approach will be adaptive to various user activity scenarios; i.e. for fixed system matrix A and given SNR the approach does not take into account the information on the transmitted signal x. Whenever a fixed system (A, SNR) receives a distorted or noisy transmitted signal y the online method tries to find a parameter α such that the reconstruction of the currently sent original signal x is possible.

The C-curve method is closely related to the known L-curve criterion that has been published by Hansen in [103, 104] and which will be in the following explained shortly. Hansen analyzed the classical ℓ_2-penalized Tikhonov functional

$$T_\alpha^2(x) := \frac{1}{2}\|Ax - y\|_2^2 + \alpha\|x\|_2^2, \qquad (8.11)$$

and recognized that plotting the discrepancy term $\|Ax_\alpha - y\|_2$ against the fitting term $\|x_\alpha\|_2$ for several solutions $x_\alpha = x(\alpha, n)$ often results in a curve $(\|Ax_\alpha - y\|_2, \|x_\alpha\|_2)$ that has the shape of a L. If the discrepancy term is plotted on the x-axis, the fitting term on the y-axis and if one chooses different regularization parameters α, one observes that a small α in (8.11) leads to an *overfitting* to the noisy measurements (i.e. $\|Ax_\alpha - y\|_2$ is quite small) and a less smooth solution x_α as $\|x_\alpha\|_2$ is potentially large. On the other hand, a very large α penalizes the fitting term much so that $\|x_\alpha\|_2$ is small, but also leads to a large discrepancy term (*under-fitting*). This simple study motivates the choice of α in a trade-off between both terms of the ℓ_2 Tikhonov functional (8.11). More precisely, Hansen has shown that it is reasonable to choose that α which corresponds to the largest curvature in the L-curve or which is closest to the coordinate origin [103].

The preceding statement motivates the transfer of the L-curve to the functionals (8.8) and (8.9), respectively. Then, one plots the discrepancy term $\|Ax_\alpha - y\|_2$ against the ℓ_1-fitting term $\|x_\alpha\|_1$, as it has already been done successfully in [147] in the application of electrical impedance tomography. In light of the given application on MUD in sporadic communication and the CDMA system setup introduced in Section 8.5, this has been done in Figure 8.2 for different SNR and

$$\alpha \in \{0, 0.001, 0.005, 0.01, 0.05, 0.1, 0.5, 1\}.$$

From top to bottom the different noise levels are treated. The left column presents the solution x_α, calculated by the RFSS algorithm, before (red dashed line) and after (blue solid line) the quantization step (8.10). The right column shows the support error $\||x| - |\hat{x}_\alpha|\|_0$ for the corresponding SNR and for all different α after quantization[3]. This measure is reasonable due to $x \in \mathcal{A}_0^{KL}$ and the fact that x_α is the projection onto the augmented alphabet $\mathcal{A}_0^{KL}$. Note that it is in fact only an error with respect to the support and does not differentiate between the symbols -1 and $+1$. The reason to choose this measure lies in the fact that also a least-squares solution would contain such errors; but there are ways, e.g. via error-correcting codes [123] to detect possibly wrong reconstructed symbols. It is therefore common in the communication community to focus on the support error only.

In Figure 8.2, it is observable that a simple plot of the discrepancy term against the fitting term of the non-quantized solution (red dashed line) does not lead to a curve that has the shape of a letter L and a parameter choice based on this curve is not obvious. If, however, the quantized solution (blue solid line) is plotted, then one recognizes for all noise levels a curve that has the shape of a C. The shape arises due to the quantization (8.10) of x_α to the alphabet $\mathcal{A}_0^{KL}$. For large α one can expect that both parts of the L- and C-curve are almost the same, as

---

[3]For both, the left and the right column in Figure 8.2, due to the stochastic properties of the setup, a reconstruction via RFSS has been calculated for 1 to 20 active users while keeping the system matrix A fixed. Then, the mean norm for the discrepancy and the fitting term has been calculated for each tuple (SNR,α).

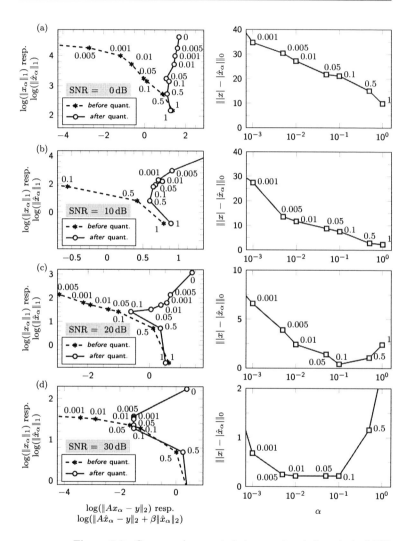

Figure 8.2.: C-curve - An a posteriori parameter choice rule for MUD problems. Each row presents for different SNR ((a) 0 dB, (b) 10 dB, (c) 20 dB, (d) 30 dB) the solution x_α based on the RFSS algorithm before and after the quantization (8.10) (which leads to $\hat{x}_\alpha$) as well as the curves describing the mean support error $\|\,|x| - |\hat{x}_\alpha|\,\|_0$ based on 1-20 active users (cf. footnote on previous page).

the solutions x_α are already very sparse and a quantization or projection step to $\hat{x}_\alpha$ only slightly effects the result. In contrast, for small α it is clear that the solution x_α tends to be more noisy. In that case, the quantization step (8.10) has a strong impact on the solution x_α which additionally implies an increased discrepancy term $\|Ax_\alpha - y\|_2$ and also causes a bending of the curve. Together with the support error curves in the right column in Figure 8.2 is it visible that for SNR > 10 one should choose that α, which is at most left in the C-curve or in other words, which minimizes the discrepancy term.

8.7. An offline approach

Another possible way to choose the regularization parameter α can be build on the knowledge on the system setting, i.e. on the system matrix A as well as on the SNR. If the system is fixed, one could set α a priori or *offline* based on a test range of a relevant number of active users that transmit a known signal x. The selection is then done by taking that parameter which minimizes the support error $\|\,|x| - |x_\alpha|\,\|_0$ with respect to all considered activity scenarios. This calibration of a fixed system has the advantage over the online method that one does not need to calculate several results x_α for different α. Instead, it suffices to choose α from a look-up table. It is, however, clear that this parameter choice rule is not insensitive to changes in the system such as the online method is.

An *oracle* or best, but also impractical selection of α in the elastic-net functional would be of course to always select exactly that parameter that considers each transmitted signal x itself for one noise level. The above described offline strategy can be seen as an approximation to the oracle approach as it tries to cover a wide range of possible transmitted signals.

8.8. Conclusion

In this chapter, the elastic-net functional has been considered for the reconstruction of signals which transmitted only sporadically from a number of nodes. For the relevant regularization parameter α in the elastic-net functional (8.9), a new online parameter choice rule was presented which was called *C-curve method*. For the numerical solution the RFSS

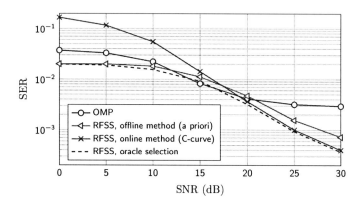

Figure 8.3.: SER curves for the OMP and RFSS variants. Except for a medium noise level of 15 dB, the RFSS outperforms the OMP. Moreover, the offline method and the online C-curve method are both near optimal for small and large SNR, respectively.

algorithm was used. In addition, an offline method was discussed to allow a quick setup based parameter choice from an a priori calculated look-up table.

A qualitative comparison of both approaches with the prominent OMP algorithm (cf. Section 6.2) in terms of the *symbol error rate* (SER) is shown in Figure 8.3 for different SNR and averaged activity scenarios as before. The SER describes the relation between the number of false reconstructed bits in in x_α compared to the true signal x with respect to the number of the overall sent bits, i.e.

$$\text{SER} = \frac{\|\,|x| - |x_\alpha|\,\|_0}{KL}.$$

The smaller the SER, the smaller the support error and the better the reconstruction performance.

What is visible in Figure 8.3 is that in comparison with the online C-curve approach, the offline method has better SER values for smaller SNR. This is a consequence of the offline approach as the parameter choice is done on the average over many instances. In contrast, the C-curve method chooses α too often wrong. This observation changes for larger SNR, where it benefits from its adaptivity to each transmitted signal rather than to its statistics. With respect to the oracle selection curve it can be seen that for small SNR, the offline method is near optimal whereas this is the case for large SNR for the C-curve method. Except for a medium noise level of around 15 dB, the RFSS outperforms the OMP. The special observation at 15 dB is probably an artifact from the fixed setting of tested regularization parameters for which the RFSS does not find a proper α.

9 | Conclusion

This thesis contributes to the fields of data compression and compressed sensing (CS) with imaging mass spectrometry (IMS) as the main application. The motivation is that IMS data is typically very large which makes the analysis for compression techniques necessary. In addition, the problem of communications technology is treated for multi-user detection in sporadic communication.

What all described models have in common is that a linear equation $Ax = y$ forms the basis for all further considerations. In compression, the matrix A was a certain dictionary Ψ in which the full given data y was assumed to be sparse. In CS, one aims in applying the compression during the acquisition process. This leads a sensing matrix $A = \Phi$ with much less rows than columns and therefore to a highly underdetermined system.

In Chapter 6, sparsity aspects of IMS data were discussed. The resulting compression techniques comprised peak picking in spectral dimension as well as smoothing of m/z-images in spatial dimension. For the spectral dimension, a new and more sensitive approach than OMP has been presented which is based on a measure for the structuredness of an m/z-image. The method and its numerical analysis were published in [8].

After the description of known image smoothing algorithms, the topic of nonnegative matrix factorization (NMF) was shortly investigated. Ongoing research results from [157] have shown that NMF has great potential to reduce the large amount of IMS data considerably with meaningful feature extraction.

Chapter 7 constitutes the main part of this thesis. There a model, which was the first of its kind, that allows CS setup in IMS was developed. In addition, it combines two typical post-processing steps of peak picking of the spectra and denoising of the m/z-images. It is proven that under certain assumptions the recovery of the data via

$$\operatorname*{argmin}_{\Lambda \in \mathbb{R}^{c \times n}} \|\Lambda\|_1 + \sum_{i=1}^{c} \|\Lambda_i\|_{TV}$$

$$\text{subject to} \quad \|Y - \mathcal{D}_{\Phi,\Psi}\Lambda\|_F \leqslant \varepsilon, \ \Lambda \geqslant 0.$$

$$(9.1)$$

is robust. The corresponding theorem is the first of its kind in CS literature as it combines the ℓ_1- and the TV-sparse perspective on the data. Moreover, the general model setup is multidimensional in the sense that it takes into account both the sparsity in the spectral dimension *and* in the spatial dimension. To the best knowledge of the author of this thesis, guarantees have so far only be considered with respect to the sparsity in the spatial dimensions in hyperspectral data [99, 100, 150]. In addition, the theorem is not restricted to orthonormal basis. Instead, any frame or overcomplete dictionary can be taken. In case of an invertible dictionary, the impact on the required number of measurements m is given by its condition number. The results were published in [18, 19].

In further research, the cited RIP result from Eftekhari *et al.* [84] for block-diagonal matrices could be extended to non-orthogonal matrices to give a more exact bound on the number of measurements required. In addition, it would be interesting to combine the NMF approach with the proposed CS model: Instead of reconstructing the full (feature) dataset $\Lambda \in \mathbb{R}_+^{c \times n}$, one would aim to extract its $\rho \ll \min\{n, c\}$ most characteristic fingerprint spectra and soft segmentation maps. Formally, one would aim to reconstruct two much smaller matrices $P \in \mathbb{R}_+^{c \times \rho}$ and $S \in \mathbb{R}_+^{\rho \times n}$, such that $\Lambda \approx PS$. The optimization problem (9.1) then reads

$$\operatorname*{argmin}_{P \in \mathbb{R}_+^{c \times \rho}, S \in \mathbb{R}_+^{\rho \times n}} \|P\|_1 + \sum_{i=1}^{\rho} \|S_i\|_{TV}$$

$$\text{subject to} \quad \|Y - \mathcal{D}_{\Phi,\Psi}PS\|_F \leqslant \varepsilon, \ \Lambda \geqslant 0.$$

For this, next to a numerical analysis, one could study if it is possible to give a similar robustness result as for (9.1). At last, especially with regard to the application in IMS, the exploration of a suitable parameter choice strategy for the numerical model parameters α and β is useful.

In Chapter 8, a known model for the sporadic communication of K sensors with one aggregation point was presented. The novelty in the chapter is given in considering the elastic-net functional for reconstructing the transmitted signal as well as in a parameter choice rule which is based on the L-curve method [103]. It was numerically shown that the results outperform the OMP for reasonable SNR and that the approach is applicable in both an online and offline scenario. The results were published in [167].

Future work could investigate an analysis on the distortion matrix H which is typically unknown, but was assumed to be known in this work. Therefore, one might investigate the dependency on H to the reconstruction properties in a CS case. For this, the presented notions of the D-RIP and the A-RIP could help to find similar results as for the robustness Theorem 7.6.8.

A | Compressed sensing results for the rat kidney

Here, results for the proposed CS in IMS approach for the rat kidney dataset (cf. Section 5.3.2) are shown.

The parameters in the optimization problem (cf. (7.35), page 121)

$$\underset{\Lambda \in \mathbb{R}^{c \times n}}{\mathrm{argmin}} \ \alpha \|\Lambda\|_1 + \beta \sum_{i=1}^{c} \|\Lambda_i\|_{TV}$$
$$\text{subject to} \ \ \|Y - \mathcal{D}_{\Phi,\Psi}\Lambda\|_F \leqslant \varepsilon, \ \Lambda \geqslant 0$$

were set for each amount of measurements by hand as follows: 40%: $\alpha = 0.2$, $\beta = 0.3$, 60%: $\alpha = 4.0$, $\beta = 4.3$, 80%: $\alpha = 9.0$, $\beta = 9.3$ and 100%: $\alpha = 19.0$, $\beta = 19.3$. The noise level was set as $\varepsilon = 7 \times 10^4$ and there were 50 outer loop iterations applied in the PPXA, cf. Section 2.2.

Figure A.1 presents the original mean spectrum (black dashed line) and its reconstruction based on different number of taken measurements (blue line). In addition, a red box highlights the effect of more taken measurements. Figures A.2 and A.3 together show reconstructions of six different m/z-images based on 40%, 60%, 80% and 100%.

As it has been discussed in Section 7.7, a reconstruction based on only 20% could not be found for all set of regularization parameters tested. This is probably a result that can be interpreted using the inequality (7.34). Compared to the rat brain dataset where there were given $c = 2000$ channels, the rat kidney data comprises of $c = 10,000$ channels. The parameter c appears in the inequality on the right hand side in a quadratic log-term. The relation states that more measurements m are needed to guaranty that the measurement matrix fulfills the RIP and thus, robust recovery of the data is possible.

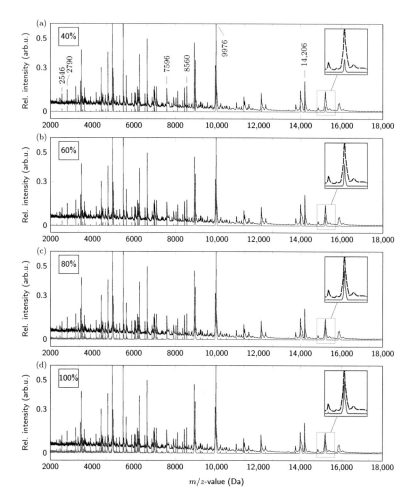

Figure A.1.: Original mean spectrum (black dashed line) and its reconstruction based on different number of taken measurements (blue line). The spectra are both normalized to $[0, 1]$ and the upper half is leaving out for better visualization. (a) Reconstruction based on 40% taken compressed measurements. (b), (c) and (d) are reconstructions based on 60%, 80% and 100%.

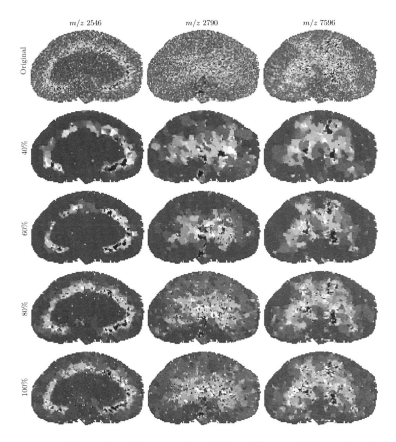

Figure A.2.: Reconstructions of three different m/z-images based on 40%, 60%, 80% and 100% of taken measurements for the kidney dataset. First column, m/z 2546 with main structures in the middle and the lower part; second column, m/z 2790 with structures at the boundaries and small regions of high intensity pixels in the middle and bottom part; third column, m/z 7596 with high intensities at the boundary as well as in the middle and the bottom.

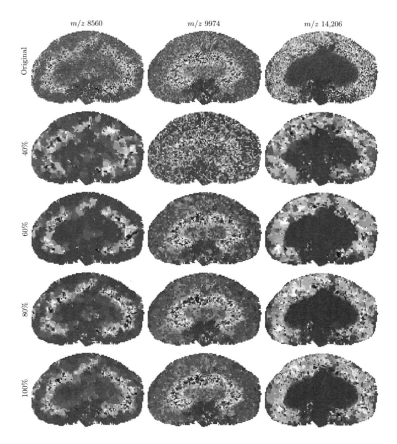

Figure A.3.: Reconstructions of three different m/z-images based on 40%, 60%, 80% and 100% of taken measurements for the kidney dataset. First column, m/z 8560 with main structures in the middle and the lower part; second column, m/z 9974 with structures at the boundaries and small regions of high intensity pixels in the middle and bottom part; third column, m/z 14,206 with high intensities at the boundary as well as in the middle and the bottom.

B | Spatial peak picking results for the rat kidney

Here results for the proposed spatial peak picking approach in imaging mass spectrometry (cf. Section 6.2) for the rat kidney dataset described in Section 5.3.2 are shown.

Figure B.1 presents the results for the detection of unknown molecules in the rat kidney dataset. The measure of chaos $\mathcal{C}$ is calculated for each m/z-image following the steps described in Section 6.2. Sorting their values in ascending order leads to the graph shown in Figure B.1(a). Figure B.1(b) shows the rat kidney data mean spectrum and the first 10, 50 and 100 selected m/z-images with lowest measure of chaos. The panels B.1(c)–(l) show m/z-images corresponding to the equidistant values of the measure pointed out in the panel (a) with their m/z-values.

It is visible that, in comparison with the results in Section 6.2.2 concerning the rat brain, the first eight images with low values of the measure of chaos look structured while the last two m/z-images (k) and (l) appear to be unstructured. This again validates the proposed method for spatial structure detection.

What can also be seen is that the m/z-image in (j) looks intuitively more structured than the one in (g) or (h); it should therefore probably has a smaller measure of chaos. However, the proposed method calculates other values. This phenomenon should be analyzed in the further research.

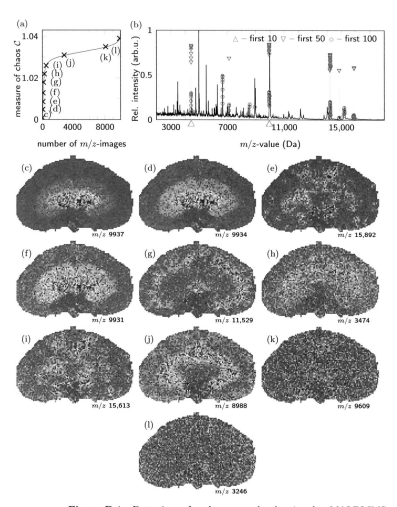

Figure B.1.: Detection of unknown molecules in the MALDI-IMS dataset from the rat kidney section. (a) Sorted values of the measure of spatial chaos for all m/z-images. Crosses indicate ten equidistant values of the measure of chaos; their m/z-images are shown in panels (c)–(l). (b) The dataset mean spectrum with the most 10 (green triangle), 50 (blue triangle), and 100 (red circle) m/z-images having lowest values of the measure of chaos. (c)–(l). m/z-images corresponding to the equidistant values of the measure pointed out in the panel (a) with their m/z-values.

Bibliography

[1] ACAR, R. and VOGEL, C. R.: Analysis of bounded variation penalty methods for ill-posed problems. *Inverse Problems*, 10(6):1217–1229, 2011.

[2] ACHLIOPTAS, D.: Database-friendly random projections. *Proceedings of the 20th ACM SIGMOD-SIGACT-SIGART symposium on principles of database systems*. ACM Press, 2001, pp. 274–281.

[3] ACHLIOPTAS, D.: Database-friendly random projections: Johnson-Lindenstrauss with binary coins. *Journal of Computer and System Sciences*, 66(4):671–687, 2003.

[4] ADCOCK, B., HANSEN, A. C., HERRHOLZ, E., and TESCHKE, G.: Generalized sampling, infinite-dimensional compressed sensing, and semi-random sampling for asymptotically incoherent dictionaries. Tech. rep. (NA13). Cambridge University - DAMTP, 2011.

[5] AHMED, N., NATARAJAN, T., and RAO, K. R.: Discrete cosine transform. *IEEE Transactions on Computers*, C-23(1):90–93, 1974.

[6] AIT-BELKACEM, R., SELLAMI, L., VILLARD, C., DEPAUW, E., CALLIGARIS, D., and LAFITTE, D.: Mass spectrometry imaging is moving toward drug protein co-localization. *Trends in Biotechnology*, 30(9):466–474, 2012.

[7] ALEXANDROV, T.: MALDI imaging mass spectrometry: statistical data analysis and current computational challenges. *BMC Bioinformatics*, 13(Suppl. 16):S11, 2012.

[8]   ALEXANDROV, T. and BARTELS, A.: Testing for presence of known and unknown molecules in imaging mass spectrometry. *Bioinformatics*, 29(18):2335–2342, 2013.

[9]   ALEXANDROV, T., BECKER, M., DEININGER, S.-O., ERNST, G., WEHDER, L., GRASMAIR, M., EGGLING, F. von, THILE, H., and MAASS, P.: Spatial segmentation of imaging mass spectrometry data with edge-preserving image denoising and clustering. *Journal of Proteome Research*, 9(12):6535–6546, 2010.

[10] ALEXANDROV, T., DECKER, J., MERTENS, B., DEELDER, A. M., TOLLENAAR, R. A. E. M., MAASS, P., and THIELE, H.: Biomarker discovery in MALDI-TOF serum protein profiles using discrete wavelet transformation. *Bioinformatics*, 25(5):643–649, 2009.

[11] ALEXANDROV, T. and KOBARG, J. H.: Efficient spatial segmentation of large imaging mass spectrometry datasets with spatially aware clustering. *Bioinformatics*, 27(ISMB 2011):i230–i238, 2011.

[12] ANDERLE, M., ROY, S., LIN, H., BECKER, C., and JOHO, K.: Quantifying reproducibility for differential proteomics: noise analysis for protein liquid chromatography-mass spectrometry of human serum. *Bioinformatics*, 20(18):3575–3582, 2004.

[13] ANDERSON, C. A. and HINTHORNE, J. R.: Ion microprobe mass analyzer. *Science*, 175(24):853–860, 1972.

[14] ANDERSSON, M., GROSECLOSE, M. R., DEUTCH, A. Y., and CAPRIOLI, R. M.: Imaging mass spectrometry of proteins and peptides: 3D volume reconstruction. *Nature Methods*, 5(1):101–108, 2008.

[15] AUBERT, G. and KORNPROBST, P.: *Mathematical Problems in Image Processing - Partial Differential Equations and the Calculus of Variations*. Springer, second ed., 2006.

[16] BALLARD, D. H.: Generalizing the Hough transform to detect arbitrary shapes. *Pattern Recognition*, 13:111–122, 1981.

[17] BARANIUK, R. G., DAVENPORT, M., DEVORE, R., and WAKIN, M.: A simple proof of the restricted isometry property for random matrices. *Constructive Approximation*, 28(3):253–263, 2008.

[18] BARTELS, A., DÜLK, P., TREDE, D., ALEXANDROV, T., and
      MAASS, P.: Compressed sensing in imaging mass spectrometry.
      Inverse Problems, 29(12):125015 (24pp), 2013.

[19] BARTELS, A., TREDE, D., ALEXANDROV, T., and MAASS, P.:
      Hybrid regularization and sparse reconstruction of imaging mass
      spectrometry data. *Proceedings of the 10th international confer-
      ence on Sampling Theory and Applications (SampTA)*. EURASIP,
      Bremen, Germany, 2013, pp. 189–192.

[20] BARTOLI, A., DOHLER, M., HERNÁNDEZ-SERRANO, J., KOUN-
      TOURIS, A., and BARTHEL, D.: Low-power low-rate goes long-
      range: the case for secure and cooperative machine-to-machine
      communications. *Proceedings of the IFIP TC 6th International
      Conference on Networking*. Springer, 2011, pp. 219–230.

[21] BAUSCHKE, H. H. and COMBETTES, P. L.: *Convex Analysis and
      Monotone Operator Theory in Hilbert Spaces*. Springer, 2011.

[22] BECK, A. and TEBOULLE, M.: A fast iterative shrinkage-
      thresholding algorithm for linear inverse problems. *SIAM Journal
      on Imaging Sciences*, 2(1):183–202, 2009.

[23] BECK, A. and TEBOULLE, M.: Fast gradient-based algorithms for
      constrained total variation image denoising and deblurring prob-
      lems. *IEEE Transactions on Image Processing*, 18(11):2419–2434,
      2009.

[24] BECKER, S., BOBIN, J., and CANDÈS, E. J.: NESTA: A fast and
      accurate first-order method for sparse recovery. *SIAM Journal on
      Imaging Sciences*, 4(1):1–39, 2011.

[25] BEHRMANN, J.: Blind source separation für MALDI-imaging.
      Bachelor thesis. University of Bremen, 2013.

[26] BLANCHARD, J. D., CARTIS, C., and TANNER, J.: Compressed
      sensing: how sharp is the restricted isometry property? *SIAM Re-
      view*, 53(1):105–125, 2011.

[27] BLANCHARD, J. D., CERMAK, M., HANLE, D., and JING, Y.:
      Greedy algorithms for joint sparse recovery. *IEEE Transactions
      on Signal Processing*, 62(7):1694–1704, 2014.

[28] BLUMENSATH, T. and DAVIES, M. E.: Iterative thresholding for sparse approximations. *Journal of Fourier Analysis and Applications*, 14(5):629–654, 2008.

[29] BLUMENSATH, T. and DAVIES, M. E.: Iterative hard thresholding for compressed sensing. *Applied and Computational Harmonic Analysis*, 27(3):265–274, 2009.

[30] BOCKELMANN, C., SCHEPKER, H. F., and DEKORSY, A.: Compressive sensing based multi-user detection for machine-to-machine communication. *Transactions on Emerging Telecommunications Technologies*, 24(4):389–400, 2013.

[31] BOGDAN, R., HANSEN, A. C., and ADCOCK, B.: On asymptotic structure in compressed sensing, 2014. arXiv: 1406.4178v2.

[32] BONESKY, T.: Morozov's discrepancy principle and Tikhonov-type functionals. *Inverse Problems*, 25(1):015015 (11pp), 2009.

[33] BONESKY, T., BREDIES, K., LORENZ, D. A., and MAASS, P.: A generalized conditional gradient method for nonlinear operator equations with sparsity constraints. *Inverse Problems*, 23(5):2041–2058, 2007.

[34] BREDIES, K. and LORENZ, D. A.: Iterated hard shrinkage for minimization problems with sparsity constraints. *SIAM Journal on Scientific Computing*, 30(2):657–683, 2008.

[35] BREDIES, K. and LORENZ, D. A.: Linear convergence of iterative soft-thresholding. *Journal of Fourier Analysis and Applications*, 14(5), 2008.

[36] BREDIES, K. and LORENZ, D. A.: *Mathematische Bildverarbeitung: Einführung in Grundlagen und moderner Theorie.* Vieweg+Teubner, 2011.

[37] BRYAN, K. and LEISE, T.: Making do with less: An introduction to compressed sensing. *SIAM Review*, 55(3):547–566, 2013.

[38] BURGER, M. and OSHER, S.: A Guide to the TV Zoo, *Level Set and PDE Based Reconstruction Methods in Imaging*, Lecture Notes in Mathematics, pp. 1–70. Springer, 2013.

[39] CAI, T. T., WANG, L., and XU, G.: Shifting inequality and recovery of sparse signals. *IEEE Transactions on Signal Processing*, 58(3):1300–1308, 2010.

[40] CAI, T. T., XU, G., and ZHANG, J.: On recovery of sparse signals via ℓ_1 minimization. *IEEE Transactions on Information Theory*, 55(7):3388–3397, 2009.

[41] CAI, T. T. and ZHANG, A: Compressed sensing and affine rank minimization under restricted isometry. *IEEE Transactions on Signal Processing*, 61(13):3279–3290, 2013.

[42] CANDÈS, E. J.: The restricted isometry property and its implications for compressed sensing. *Comptes Rendus Mathematique*, 346(9–10):589–592, 2008.

[43] CANDÈS, E. J., ELDAR, Y. C., NEEDELL, D., and RANDALL, P.: Compressed sensing with coherent and redundant dictionaries. *Applied and Computational Harmonic Analysis*, 31(1):59–73, 2011.

[44] CANDÈS, E. J. and ROMBERG, J.: ℓ_1-MAGIC. 2005. URL: http://users.ece.gatech.edu/~justin/l1magic/ (visited on 06/18/2014).

[45] CANDÈS, E. J. and ROMBERG, J.: Sparsity and incoherence in compressive sampling. *Inverse Problems*, 23(3):969–985, 2007.

[46] CANDÈS, E. J., ROMBERG, J., and TAO, T.: Robust uncertainty principles: exact signal reconstruction from highly incomplete frequency information. *IEEE Transactions on Information Theory*, 52(2):489–509, 2006.

[47] CANDÈS, E. J., ROMBERG, J., and TAO, T.: Stable signal recovery from incomplete and inaccurate measurements. *Communications on Pure and Applied Mathematics*, 59(8):1207–1223, 2006.

[48] CANDÈS, E. J. and TAO, T.: Decoding by linear programming. *IEEE Transactions on Information Theory*, 51(12):4203–4215, 2005.

[49] CANDÈS, E. J. and TAO, T.: Near optimal signal recovery from random projections: universal encoding strategies? *IEEE Transactions on Information Theory*, 52(12):5406–5425, 2006.

[50]   CANDÈS, E. J. and WAKIN, M. B.: An introduction to compressive sampling. *IEEE Signal Processing Magazine*, 25(2):21–30, 2008.

[51]   CAPRIOLI, R. M.: Imaging mass spectrometry: molecular microscopy for enabling a new age of discovery. *Proteomics*, 23(7–8):807–809, 2014.

[52]   CAPRIOLI, R. M., FARMER, T. B., and GILE, J.: Molecular imaging of biological samples: localization of peptides and proteins using MALDI-TOF MS. *Analytical Chemistry*, 23(69):4751–4760, 1997.

[53]   CHAMBOLLE, A.: An algorithm for total variation minimization and applications. *Journal of Mathematical Imaging and Vision*, 20(1-2):89–97, 2004.

[54]   CHAN, T. F. and VESE, L. A.: Active contours without edges. *IEEE Transactions on Image Processing*, 10(2):266–277, 2001.

[55]   CHAURAND, P.: Imaging mass spectrometry of thin tissue sections: a decade of collective efforts. *Journal of Proteomics*, 16(75):4883–4892, 2012.

[56]   CHEN, S. S., DONOHO, D. L., and SAUNDERS, M. A.: Atomic decomposition by basis pursuit. *SIAM Journal on Scientific Computing*, 20(1):33–61, 1998.

[57]   CLASON, C., JIN, B., and KUNISCH, K.: A semismooth Newton method for ℓ_1 data fitting with automatic choice of regularization parameters and noise calibration. *SIAM Journal on Imaging Sciences*, 3(2):199–231, 2010.

[58]   COHEN, A., DAHMEN, W., and DEVORE, R. A.: Compressed sensing and best k-term approximation. *Journal of the American Mathematical Society*, 22(1):211–231, 2009.

[59]   COMBETTES, P. L. and PESQUET, J.-C.: A proximal decomposition method for solving convex variational inverse problems. *Inverse problems*, 24(6):27, 2008.

[60]   COMBETTES, P. L. and PESQUET, J.-C.: *Proximal splitting methods in signal processing*. Vol. 49 of *Fixed-Point Algorithms for Inverse Problems in Science and Engineering*. Springer, 2011, pp. 185–212.

[61] CONSORTIUM, EXALTED: D2.1 - description of baseline reference systems, scenarios, technical requirements & evaluation methodology. Tech. rep. European Collaborative Research FP7 - Project No. 258512, May 2011.

[62] COOMBES, K. R., FRITSCHE, H. A., CLARKE, C., CHEN, J.-N., BAGGERLY, K. A., MORRIS, J. S., XIAO, L.-C., HUNG, M.-C., and KUERER, H. M.: Quality control and peak finding for proteomics data collected from nipple aspirate fluid by surface-enhanced laser desorption and ionization. *Clinical Chemistry*, 49(10):1615–1623, 2003.

[63] COOMBES, K. R., KOOMEN, J. M., BAGGERLY, K. A., MORRIS, J. S., and KOBAYASHI, R.: Understanding the characteristics of mass spectrometry data through the use of simulation. *Cancer Informatics*, 1(1):41–52, 2005.

[64] CORMEN, T. H., LEISERSON, C. E., and RIVEST, R. L.: *Introduction to Algorithms*. MIT Press, second ed., 2001.

[65] DAUBECHIES, I.: *Ten Lectures on Wavelets*. CBMS-NSF Regional Conference Series in Applied Mathematics, first ed., 1992.

[66] DAUBECHIES, I., DEFRISE, M., and DE MOL, C.: An iterative thresholding algorithm for linear inverse problems with a sparsity constraint. *Communications on Pure and Applied Mathematics*, 57(11):1413–1457, 2004.

[67] DEININGER, S.-O., CORNETT, D. S., PAAPE, R., BECKER, M., PINEAU, C., RAUSER, S., WALCH, A., and WOLSKI, E.: Normalization in MALDI-TOF imaging datasets of proteins: practical considerations. *Analytical and Bioanalytical Chemistry*, 401(1):167–181, 2011.

[68] DENIS, L., LORENZ, D. A., and TREDE, D.: Greedy solution of ill-posed problems: error bounds and exact inversion. *Inverse Problems*, 25(11):115017 (24pp), 2009.

[69] DEVORE, R. A.: Deterministic constructions of compressed sensing matrices. *Journal of Complexity*, 23(4):918–925, 2007.

[70] DI MARCO, V. B. and BOMBI, G. G.: Mathematical functions for the representation of chromatographic peaks. *Journal of Chromatography A*, 931(1–2):1–30, 2001.

[71] DONOHO, D. L.: Sparse components of images and optimal atomic decompositions. *Constructive Approximation*, 17(3):353–382, 2001.

[72] DONOHO, D. L.: Compressed sensing. *IEEE Transactions on Information Theory*, 52(4):1289–1306, 2006.

[73] DONOHO, D. L. and ELAD, M.: Optimally sparse representation in general (nonorthogonal) dictionaries via ℓ_1 minimization. *PNAS*, 100(5):2197–2202, 2003.

[74] DONOHO, D. L. and STARK, P. B.: Uncertainty principles and signal recovery. *SIAM Journal on Applied Mathematics*, 49(3):906–931, 1989.

[75] DONOHO, D. L. and STODDEN, V.: When does non-negative matrix factorization give a correct decomposition into parts? *Advances in Neural Information Processing Systems (NIPS) 16*. MIT Press, 2003.

[76] DONOHO, D. L. and STODDEN, V.: Breakdown point of model selection when the number of variables exceeds the number of observations. *Proceedings of the International Joint Conference on Neural Networks*, 2006.

[77] DONOHO, D. L. and TANNER, J.: Observed universality of phase transitions in high-dimensional geometry, with implications for modern data analysis and signal processing. *Philosophical Transactions of the Royal Society A*, 367(1906):4273–4293, 2009.

[78] DUARTE, M. F. and BARANIUK, R. G.: Kronecker compressive sensing. *IEEE Transactions on Image Processing*, 21(2):494–504, 2012.

[79] DUARTE, M. F., DAVENPORT, M. A., TAKHAR, D., LASKA, J. N., SUN, T., KELLY, K. F., and BARANIUK, R. G.: Single-pixel imaging via compressive sampling. *IEEE Signal Processing Magazine*, 25(2):83–91, 2008.

[80] DUARTE, M. F. and ELDAR, Y. C.: Structured compressed sensing: from theory to applications. *IEEE Transactions on Signal Processing*, 59(9):4053–4085, 2011.

[81] DUDA, R. O., HART, P. E., and STORK, D. G.: *Pattern Classification*. Wiley, second ed., 2001.

[82] ECKSTEIN, J. and BERTSEKAS, D. P.: On the Douglas-Rachford splitting method and the proximal point algorithm for maximal monotone operators. *Mathematical Programming*, 55(3):293–318, 1992.

[83] EDELMAN, A. and RAO, N. R.: Random matrix theory. *Acta Numerica*, 14:233–297, May 2005.

[84] EFTEKHARI, A., YAP, H. L., ROZELL, C., and WAKIN, M. B.: The restricted isometry property for random block diagonal matrices. *Applied and Computational Harmonic Analysis*, 2014. accepted.

[85] ELAD, M., MILANFAR, P., and RUBINSTEIN, R.: Analysis versus synthesis in signal priors. *Inverse Problems*, 23(3):947–968, 2007.

[86] ELDAR, Y. C. and KUTYNIOK, G.: *Compressed Sensing - Theory and Applications*. Cambridge University Press, 2012.

[87] ENGL, H. W., HANKE, M., and NEUBAUER, A.: *Regularization of Inverse Problems*. Springer Netherlands, 2000.

[88] FADILI, M.-J. and STARCK, J.-L.: Monotone operator splitting for optimization problems in sparse recovery. *ICIP 2009*:1461–1464, 2009.

[89] FALLAT, S. M. and JOHNSON, C. R.: *Totally Nonengative Matrices*. Princeton Series in Applied Mathematics, first ed., 2011.

[90] FIGUEIREDO, M. A. T., NOWAK, R. D., and WRIGHT, S. J.: Gradient projection for sparse reconstruction: Application to compressed sensing and other inverse problems. *IEEE Journal of Selected Topics in Signal Processing*, 1(4):586–597, 2007.

[91] FOLEY, J. P.: Equations for chromatographic peak modeling and calculation of peak area. *Analytical Chemistry*, 59(15):1984–1987, 1987.

[92] FOUCART, S. and LAI, M.-J.: Sparsest solutions of underdetermined linear systems via ℓ_q-minimization for $0 < q \leqslant 1$. *Applied and Computational Harmonic Analysis*, 26(3):395–407, 2009.

[93] FOUCART, S. and RAUHUT, H.: *A Mathematical Introduction to Compressive Sensing*. Springer, 2014.

[94] FYHN, K., JENSEN, T. L., LARSEN, T., and JENSEN, S. H.: Compressive sensing for spread spectrum receivers. *IEEE Transactions on Wireless Commonications*, 12(5):2334–2343, 2013.

[95] GAO, Y., ZHU, L., NORTON, I., AGAR, N. Y. R., and TANNENBAUM, A.: Reconstruction and feature selection for desorption electrospray ionization mass spectroscopy imagery. *Proceedings: SPIE, Medical Imaging: Image-Guided Procedures, Robotic Interventions, and Modeling*. Vol. 9063. SPIE, 2014.

[96] GAUTSCHI, W.: Norm estimates for inverses of Vandermonde matrices. *Numerische Mathematik*, 23:337–347, 1974/75.

[97] GOLBABAEE, M., ARBERET, S., and VANDERGHEYNST, P.: Distributed compressed sensing of hyper-spec-tral images via blind source separation. *Presentation given at Asilomar Conference on signals, systems, and computers, Pacific Groove, CA, USA, November 7-10*, 2010.

[98] GOLBABAEE, M., ARBERET, S., and VANDERGHEYNST, P.: Multichannel compressed sensing via source separation for hyperspectral images. *Eusipco 2010, Aalborg, Denmark*, 2010.

[99] GOLBABAEE, M., ARBERET, S., and VANDERGHEYNST, P.: Compressive source separation: theory and methods for hyperspectral imaging. *IEEE Transactions on Image Processing*, 22(12):5096–5110, 2013.

[100] GOLBABAEE, M. and VANDERGHEYNST, P.: Joint trace/TV norm minimization: a new efficient approach for spectral compressive imaging, 2012.

[101] GRASMAIR, M.: Locally adaptive total variation regularization, *Scale Space and Variational Methods in Computer Vision*. Vol. 5567, Lecture Notes in Computer Science, pp. 331–342. Springer, 2009.

[102] GRIESSE, R. and LORENZ, D. A.: A semismooth Newton method for Tikhonov functionals with sparsity constraints. *Inverse Problems*, 24(3):035007 (19pp), 2008.

[103] HANSEN, P. C.: Analysis of discrete ill-posed problems by means of the L-curve. *SIAM Review*, 34(4):561–580, 1992.

[104] HANSEN, P. C. and O'LEARY, D. P.: The use of the L-curve in the regularization of discrete ill-posed problems. *SIAM Journal on Scientific Computing*, 14(6):1487–1503, 1993.

[105] HERRHOLZ, E., LORENZ, D. A., TESCHKE, G., and TREDE, D.: *Sparsity and compressed sensing in inverse problems.* to appear as book chapter, 2014.

[106] HOLLE, A., HAASE, A., KYSER, M., and HÖNDORF, J.: Optimizing UV laser focus profiles for improved MALDI performance. *Journal of Mass Spectrometry*, 41(6):705–716, 2006.

[107] HORN, R. A. and JOHNSON, C. R.: *Topics in Matrix Analysis.* Cambridge University Press, first ed., 1991.

[108] HOYER, P. O. and DAYAN, P.: Non-negative matrix factorization with sparseness constraints. *Journal of Machine Learning Research*, 5:1457–1469, 2004.

[109] JIN, B., LORENZ, D. A., and SCHIFFLER, S.: Elastic-Net Regularization: Error estimates and Active Set Methods. *Inverse problems*, 25(11):115022 (26pp), 2009.

[110] JOKAR, S. and MEHRMANN, V.: Sparse solutions to underdetermined Kronecker product systems. *Linear algebra and its applications*, 431(12):2437–2447, 2009.

[111] JONES, E. A., SHYTI, R., ZEIJL, R. J. M. van, HEININGEN, S. H. van, FERRARI, M. D., DEELDER, A. M., TOLNER, E. A., MAAG-DENBERG, A. M.J.M. van den, and MCDONNELL, L. A.: Imaging mass spectrometry to visualize biomolecule distributions in mouse brain tissue following hemispheric cortical spreading depression. *Journal of Proteomics*, 75(16):5027–5035, 2012.

[112] KAMMEYER, K. D.: *Nachrichtentechnik.* Vieweg+Teubner, fifth ed., 2011.

[113] KARAS, M. and HILLENKAMP, F.: Laser desorption ionization of proteins with molecular masses exceeding 10,000 daltons. *Analytical Chemistry*, 60(20), 1988.

[114] KLERK, L. A., BROERSEN, A., FLETCHER, I. W., LIERE, R. van, and HEEREN, R. M. A.: Extended data analysis strategies for high resolution imaging MS: New methods to deal with extremely large image hyperspectral datasets. *International Journal of Mass Spectrometry*, 260(2-3):222–236, 2007.

[115] KOBARG, J. H.: Signal and image processing methods for imaging mass spectrometry data. PhD thesis. University of Bremen, 2014.

[116] KOBARG, J. H., MAASS, P., OETJEN, J., TROPP, O., HIRSCH, E., SAGIV, C., GOLBABAEE, M., and VANDERGHEYNST, P.: Numerical experiments with MALDI imaging data. *Advances in Computational Mathematics*:1–16, 2013.

[117] KOHAVI, R.: A study of cross-validation and bootstrap for accuracy estimation and model selection. *Proceedings of the 14th International Joint Conference on Artificial Intelligence*. Vol. 2. IJCAI'95, 1995, pp. 1137–1143.

[118] KRAHMER, F. and WARD, R.: Stable and robust sampling strategies for compressive imaging. *IEEE Transactions on Image Processing*, 23(2):612–622, 2014.

[119] LABATE, D., LIM, W.-Q., KUTYNIOK, G., and WEISS, G.: Sparse multidimensional representation using shearlets. *Wavelets XI - Proceedings of the SPIE*, 2005, pp. 254–262.

[120] LEE, D. D. and SEUNG, H. S.: Learning the parts of objects by non-negative matrix factorization. *Nature*, 401(6755):788–791, 1999.

[121] LEE, D. D. and SEUNG, H. S.: Algorithms for non-negative matrix factorization. *Advances in Neural Information Processing Systems (NIPS)*. MIT Press, 2000, pp. 556–562.

[122] LEE, H., BATTLE, A., RAINA, R., and NG, A. Y.: Efficient sparse coding algorithms. SCHÖLKOPF, B., PLATT, J., and HOFFMAN, T., editors, *Advances in Neural Information Processing Systems 19*, pp. 801–808. MIT Press, 2007.

[123] LIN, S. and COSTELLO, D. J.: *Error Control Coding: Fundamentals and Applications*. Pearson-Prentice Hall, second ed., 2004.

[124] LIU, Y., LI, S., MI, T., LEI, H., and YU, W.: Performance analysis of ℓ_1-synthesis with coherent frames. *Proceedings: IEEE International Symposium on Information Theory Proceedings (ISIT)*, 2012, pp. 2042–2046.

[125] LIU, Y., MI, T., and LI, S.: Compressed sensing with general frames via optimal-dual-based ℓ_1-analysis. *IEEE Transactions on Information Theory*, 58(7):4201–4214, 2012.

[126] LORENZ, D. A.: Convergence rates and source conditions for Tikhonov regularization with sparsity constraints. *Journal of Inverse and Ill-posed Problems*, 16(5):463–478, 2008.

[127] LORENZ, D. A. and POCK, T.: An accelerated forward-backward algorithm for monotone inclusions. submitted. 2014.

[128] LORIS, I.: On the performance of algorithms for the minimization of ℓ_1-penalized functionals. *Inverse Problems*, 25(3):035008 (16pp), 2009.

[129] LOUIS, A. K.: *Inverse und schlecht gestellte Probleme*. Vieweg+Teubner, 1989.

[130] LOUIS, A. K.: Feature reconstruction in inverse problems. *Inverse Problems*, 27(6):1–21, 2011.

[131] LOUIS, A. K., MAASS, P., and RIEDER, A.: *Wavelets: Theory and Applications*. Vieweg+Teubner, first ed., 1994.

[132] LUSTIG, M., DONOHO, D. L., and PAULY, J. M.: Sparse MRI: the application of compressed sensing for rapid MR imaging. *Magnetic Resonance in Medicine*, 58(6):1182–1195, 2007.

[133] LUSTIG, M., DONOHO, D. L., SANTOS, J. M., and PAULY, J. M.: Compressed sensing MRI. *IEEE Signal Processing Magazine*, 25(2):72–82, 2008.

[134] MAASS, P. and STARK, H. G.: Wavelets and digital image processing. *Surveys on Mathematics for Industry*, 4(3):195–235, 1994.

[135] MAIRAL, J.: Optimization with first-order surrogate functions. *Proceedings of the International Conference on Machine Learning (ICML)*, 2013.

[136] MALLAT, S. G.: *A Wavelet Tour of Signal Processing*. Academic Press, third ed., 2008.

[137] MALLAT, S. G., DAVIS, G., and ZHANG, Z.: Adaptive time-frequency decompositions. *SPIE Optical Engineering*, 3(7):2183–2191, 1994.

[138] MALLAT, S. G. and ZHANG, Z.: Matching pursuits with time-frequency dictionaries. *IEEE Transactions on Signal Processing*, 41(12):3397–3415, 1993.

[139] MARKIDES, K. and GRÄSLUND, A.: Mass spectrometry (MS) and nuclear magnetic resonance (NMR) applied to biological macro-molecules. Advanced information on the nobel prize in chemistry 2002. Tech. rep. The Royal Swedish Academy of Science, 2002.

[140] MATLAB: *Version 8.1.0.604 (R2013a)*. The MathWorks Inc., Natick, Massachusetts, 2013.

[141] MCCOMBIE, G., STAAB, D., STOECKLI, M., and KNOCHENMUSS, R.: Spatial and spectral correlations in MALDI mass spectrometry images by clustering and multivariate analysis. *Analytical Chemistry*, 77(19):6118–6124, 2005.

[142] MCDONNELL, L. A., REMOORTERE, A. van, ZEIJL, R. J. M. van, and DEELDER, A. M.: Mass spectrometry image correlation: quantifying colocalization. *Journal of Proteome Research*, 7(8):3619–3627, 2008.

[143] MOHIMANI, H., BABAIE-ZADEH, B., and JUTTEN, C.: A fast approach for overcomplete sparse decomposition based on smoothed ℓ^0 norm. *IEEE Transactions on Signal Processing*, 57(1):289–301, 2009.

[144] MONDAL, T., JAIN, A., and SARDANA, H. K.: Automatic craniofacial structure detection on cephalometric images. *IEEE Transactions on Image Processing*, 20:2606–2614, 2011.

[145] MOREAU, J.-J.: Proximité et dualité dans un espace hilbertian. *Bulletin de la Société Mathématique de France*, 93:273–299, 1965.

[146] MOROZOV, V. A.: On the solution of functional equations by the method of regularization. *Soviet Mathematics - Doklady*, 7:414–417, 1966.

[147] NASEHI TEHRANI, J., MCEWAN, A., JIN, C., and SCHAIK, A. van: L1 regularization method in electrical impedance tomography by using the L1-curve (Pareto frontier curve). *Applied Mathematical Modelling*, 36(3):1095–1105, 2012.

[148] NATARAJAN, B.: Sparse approximation solutions to linear systems. *SIAM Journal on Computing*, 24(2):227–234, 1995.

[149] NEEDELL, D. and TROPP, J. A.: CoSaMP: Iterative signal recovery from incomplete and inaccurate samples. *Applied and Computational Harmonic Analysis*, 26(3):301–321, 2009.

[150] NEEDELL, D. and WARD, R.: Near-optimal compressed sensing guarantees for total variation minimization. *IEEE Transactions on Image Processing*, 22(10):3941–3949, 2013.

[151] NEEDELL, D. and WARD, R.: Stable image reconstruction using total variation minimization. *SIAM Journal on Imaging Sciences*, 6(2):1035–1058, 2013.

[152] NYQUIST, H.: Certain topics in telegraph transmission theory. *IEEE Transactions of the A.I.E.E.*, 47:617–644, 1928.

[153] PALMER, A. D., BUNCH, J., and STYLES, I. B.: Randomized approximation methods for the efficient compression and analysis of hyperspectral data. *Analytical Chemistry*, 85(10):5078–5086, 2013.

[154] PARIS, S., KORNPROBST, P., TUMBLIN, J., and DURAND, F.: Bilateral filtering: theory and applications. *Found. and Trends in Computer Graphics and Vision*, 4(1):1–73, 2008.

[155] PATI, Y. C., REZAIIFAR, R., and KRISHNAPRASAD, P. S.: Orthogonal matching pursuit: recursive function approximation with applications to wavelet decomposition. *Proceedings of 27th Asilomar Conference on Signals, Systems and Computers*. Vol. 1, 1993, pp. 40–44.

[156] PENNEBAKER, W. B. and MITCHELL, J. L.: *JPEG still image data compression standard*. Springer, third ed., 1993.

[157] PHAM, Q. M., BARTELS, A., and MAASS, P.: Non-smooth minimization problems of matrix variable and applications. in preparation. 2014.

[158] PIEHOWSKI, P. D., DAVEY, A. M., KURCZY, M. E., SHEETS, E. D., WINOGRAD, N., EWING, A. G., and HEIEN, M. L.: Time-of-fight secondary ion mass spectrometry imaging of subcellular lipid heterogeneity: Poisson counting and spatial resolution. *Analytical Chemistry*, 81(14):5593–5602, 2009.

[159] RABBANI, M. and JONES, P. W.: *Digital Image Compression Techniques*. SPIE Press, first ed., 1992.

[160] RAUHUT, H., SCHNASS, K., and VANDERGHEYNST, P.: Compressed sensing and redundant dictionaries. *IEEE Transactions on Information Theory*, 54(5):2210–2219, 2008.

[161] RIEDER, A.: *Keine Probleme mit Inversen Problemen*. Vieweg, 2003.

[162] ROMANO, G., PALMIERI, F., and WILLETT, P. K.: Soft iterative decoding for overloaded CDMA. *Proceedings of the IEEE International Conference on Acoustics, Speech, and Signal Processing*. Vol. 3, 2005, pp. iii/733–iii/736.

[163] RUDIN, L. I., OSHER, S., and FATEMI, E.: Nonlinear total variation based noise removal algorithms. *Physica D: Nonlinear Phenomena*, 60:259–268, 1992.

[164] SALOMON, D.: *Data Compression: The Complete Reference*. Springer, third ed., 2004.

[165] SAWATZKY, A., BRUNE, C., MÜLLER, J., and BURGER, M.: Total variation processing of images with Poisson statistics. *CAIP 2009 - Lecture Notes in Computer Science*, 5702:533–540, 2009.

[166] SAYOOD, K.: *Introduction to Data Compression*. Elsevier Science, first ed., 2012.

[167] SCHEPKER, H. F., BOCKELMANN, C., DEKORSY, A., BARTELS, A., TREDE, D., and KAZIMIERSKI, K. S.: C-curve: A finite alphabet based parameter choice rule for elastic-net in sporadic communication. *IEEE Signal Processing Letters*, 18(8):1443–1446, 2014.

[168] SCHEPKER, H., BOCKELMANN, C., and DEKORSY, A.: Coping with CDMA asynchronicity in compressive sensing multi-user detection. *Proceedings of the IEEE 77th Vehicular Technology Conf.* 2013.

[169] SCHEPKER, H. and DEKORSY, A.: Sparse multi-user detection for CDMA transmission using greedy algorithms. *Proceedings of the 8th Int. Symp. on Wireless Comm. Systems*, 2011.

[170] SCHEPKER, H. and DEKORSY, A.: Compressive sensing multi-user detection with block-wise orthogonal least squares. *Proceedings of the IEEE 75th Vehicular Technology Conf.* 2012.

[171] SCHERZER, O.: *Handbook of Mathematical Methods in Imaging.* Springer, 2010.

[172] SCHIFFLER, S.: The elastic-net: Stability for sparsity methods. PhD thesis. University of Bremen, 2010.

[173] SCHÖNE, C., HÖFLER, H., and WALCH, A. *Clinical Biochemistry*, 46(6):539–545, 2013.

[174] SEELEY, E. H. and CAPRIOLI, R. M.: 3d imaging by mass spectrometry: a new frontier. *Analytical Chemistry*, 84(5):2105–2110, 2012.

[175] SHANNON, C. E.: Communication in the presence of noise. *Proceedings of the Institute of Radio Engineers*, 37(1):10–21, 1949.

[176] SHIM, B. and SONG, B.: Multiuser detection via compressive sensing. *IEEE Communication Letters*, 16(7):972–974, 2012.

[177] STARCK, J.-L., CANDÈS, E. J., and DONOHO, D. L.: The curvelet transform for image denoising. *IEEE Transactions on Image Processing*, 11(6):670–684, 2002.

[178] STARCK, J.-L., MURTAGH, F., and FADILI, J. M.: *Sparse Image and Signal Processing - Wavelets, Curvelets, Morphological Diversity.* Cambridge, first ed., 2010.

[179] STEINHORST, Klaus: Sparsity- and graph-regularized nonnegative matrix factorization with application in life sciences. Diploma thesis. University of Bremen, 2011.

[180] STOCKHAM, T. G., CANNON, T. M., and INGEBRETSEN, R. B.: Blind deconvolution through digital signal processing. *Proceedings of the IEEE*, 63(4):678–692, 1975.

[181] STOECKLI, M., CHAURAND, P., HALLAHAN, D. E., and CAPRIOLI, R. M.: Imaging mass spectrometry: a new technology for the analysis of protein expression in mammalian tissues. *Nature Medicine*, 7(4):493–496, 2001.

[182] SUN, T. and KELLY, K. F.: Compressive sensing hyperspectral imager. *Proceedings on Computational Optical Sensing and Imaging*, 2009.

[183] TAKHAR, D., LASKA, J. N., WAKIN, M. B., DUARTE, M. F., BARON, D., SARVOTHAM, S., KELLY, K. F., and BARANIUK, R. G.: A new compressive imaging camera architecture using optical-domain compression. *Proceedings of Computational Imaging IV at SPIE Electronic Imaging*. Vol. 6065, 2006, pp. 43–52.

[184] TAUBMAN, D. S. and MARCELLIN, M.: *JPEG2000: Image Compression Fundamentals, Standards and Practice*. Kluwer International Series in Engineering and Computer Science, first ed., 2001.

[185] THE ECONOMIST: Data, data everywhere. Special Issue. Feb. 2010.

[186] TIBSHIRANI, R.: Regression shrinkage and selection via the lasso. *Journal of the Royal Statistical Society, Series B*, 58(1):267–288, 1996.

[187] TIKHONOV, A. N. and ARSENIN, V. Y.: *Solutions of ill-posed problems*. V. H. Winston & Sons, 1977.

[188] TIKHONOV, A. N. and GLASKO, V. B.: Use of the regularization method in non-linear problems. *USSR Computational Mathematics and Mathematical Physics*, 5(3):93–107, 1965.

[189] TILLMANN, A. M. and PFETSCH, M. E.: The computational complexity of the restricted isometry property, the nullspace property, and related concepts in compressed sensing. *IEEE Transactions on Information Theory*, 60(2):1248–1259, 2014.

[190] TOMASI, C. and MANDUCHI, R.: Bilateral filtering for gray and color images. *Proceedings of the 6th International Conference on Computer Vision*, 1998, pp. 839–846.

[191] TREDE, D.: *Inverse Problems with Sparsity Constraints: Convergence Rates and Exact Recovery.* Logos Berlin, 2010.

[192] TREDE, D., KOBARG, J. H., OETJEN, J., THIELE, H., MAASS, P., and ALEXANDROV, T.: On the importance of mathematical methods for analysis of MALDI-imaging mass spectrometry data. *Journal of Integrative Bioinformatics*, 9(1):189–200, 2012.

[193] TSE, D. and VISWANATH, P.: *Fundamentals of Wireless Communication.* Cambridge University Press, first ed., 2005.

[194] UCHIDA, Y.: A simple proof of the geometric-arithmetic mean inequality. *Journal of Inequalities in Pure and Applied Mathematics*, 9(2):2 pp. 2008.

[195] UNSER, M.: Sampling – 50 years after Shannon. *Proceedings of IEEE*, 88(4):569–587, 2000.

[196] WAHLBERG, B., BOYD, S., ANNERGREN, M., and WANG, Y.: An ADMM algorithm for a class of total variation regularized estimation problems. *Proc. 16th IFAC Symposium on System Identification.* Vol. 16. (1), 2012.

[197] WAKIN, M. B., LASKA, J. N., DUARTE, M. F., BARON, D., SARVOTHAM, S., TAKHAR, D., KELLY, K. F., and BARANIUK, R. G.: An architecture for compressive imaging. *Proceedings of the IEEE International Conference on Image Processing*, 2006, pp. 1273–1276.

[198] WANG, X. and POOR, H. V.: *Wireless Communication Systems: Advanced Techniques for Signal Reception.* Prentice Hall, first ed., 2003.

[199] WANG, Y., YANG, J., YIN, W., and ZHANG, Y.: A new alternating minimization algorithm for total variation image reconstruction. *SIAM Journal on Imaging Sciences*, 1(3):248–272, 2008.

[200] WANG, Z. and BOVIK, A. C.: Mean squared error: love it or leave it? a new look at signal fidelity measures. *IEEE Signal Processing Magazine*, 26(1):98–117, 2009.

[201] WANG, Z., BOVIK, A. C., SHEIKH, H. R., and SIMONCELLI, E. P.: Image quality assessment: from error visibility to structural similarity. *IEEE Transactions on Image Processing*, 13(4):600–612, 2004.

[202] WATROUS, J. D., ALEXANDROV, T., and DORRESTEIN, P. C.: The evolving field of imaging mass spectrometry and its impact on future biological research. *Journal of Mass Spectrometry*, 46(2):209–222, 2011.

[203] WATROUS, J., ROACH, P., ALEXANDROV, T., HEATH, B. S., YANG, J. Y., KERSTEN, R. D., VOORT, M. van der, POGLIANO, K., GROSS, H., RAAIJMAKERS, J. M., MOORE, B. S., LASKIN, J., BANDEIRA, N., and DORRESTEIN, P. C.: Mass spectral molecular networking of living microbial colonies. *PNAS*, 109(26):E1743–E1752, 2012.

[204] WATSON, D. S., PIETTE, M. A., SEZGEN, O., and MOTEGI, N.: Machine to machine (m2m) technology in demand responsive commercial buildings. *Proceedings of ACEEE Summer Study on Energy Efficiency in Buildings*, 2004.

[205] WATSON, J. T. and SPARKMAN, O. D.: *Introduction to Mass Spectrometry: Instrumentation, Applications, and Strategies for Data Interpretation*. Wiley, fourth ed., 2007.

[206] WELCH, L.: Lower bounds on the maximum cross correlation of signals. *IEEE Transactions on Information Theory*, 20(3):397–399, 1974.

[207] WISEMAN, J. M., IFA, D. R., SONG, Q., and COOKS, R. G.: Tissue imaging at atmospheric pressure using desorption electrospray ionization (DESI) mass spectrometry. *Angewandte Chemie Int. Ed.*, 45(43):7188–7192, 2006.

[208] ZHU, H. and GIANNAKIS, G. B.: Exploting sparse user activity in multiuser detection. *IEEE Transactions on Communications*, 59(2):454–465, 2011.

[209] ZOU, H. and HASTIE, T.: Regularization and variable selection via the elastic net. *Journal of the Royal Statistical Society, Series B*, 67:301–320, 2005.

[210] ZWEIGENBAUM, J.: *Mass Spectrometry in Food Safety - Methods and Protocols.* Of *Methods in Molecular Biology.* Springer, Humana Press, 2011.